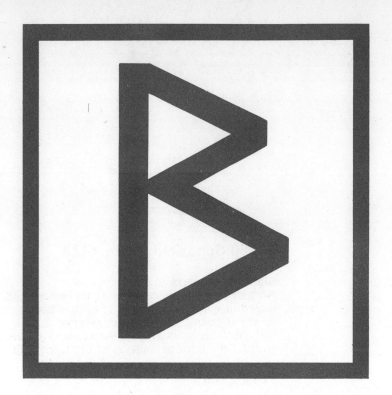

Beta Mathematics 4

Compiled by

In association with

T. R. Goddard & A. W. Grattidge

J. W. Adams & R. P. Beaumont

 Schofield & Sims Ltd Huddersfield

7217 2105 2

First printed 1969
Second impression 1969
Third impression 1969

Designed and Printed in England by Chorley & Pickersgill Ltd Leeds
Bound in Scotland

Contents   Beta Mathematics 4

Tens number system

Thousands Th	Hundreds H	Tens T	Units U	tenths t	hundredths h
a			7	2 ·	9	
b		5	0	6 ·	8	4
c	2	4	3	0 ·	0	7
d				0 ·	7	9

In example **a** the number is read as seventy-two point nine.

1 Read examples **b, c, d** to your partner in the same way.

2 Read these numbers to your partner, then draw columns like those above and write them.
   a 7·8   **b** 90·03   **c** 0·37   **d** 807·6   **e** 1 004

3 Write in words and figures the number shown in each abacus picture.

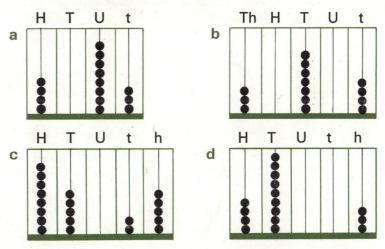

4 Write the value of the figure 5 in each of these numbers
   a 185·3   **b** 17·45   **c** 503·28   **d** 5 070.

5 Write the value of the figure 3 in each of these numbers
   a 307·92   **b** 130·6   **c** 59·34   **d** 8·03.

B

1 Add 1 to each of these numbers
   a 399   **b** 9·46   **c** 2 009   **d** 89·5.

2 Add 10 to each of these numbers
   a 790   **b** 3 990   **c** 90·7   **d** 4·76.

3 Add 100 to each of these numbers
   a 906   **b** 3 039   **c** 7 005   **d** 1·37.

4 Subtract 1 from each of these numbers
   a 300   **b** 100·4   **c** 1·92   **d** 1 010.

5 Subtract 10 from each of these numbers
   a 27·42   **b** 1 017   **c** 4 006   **d** 11·03.

6 Subtract 100 from each of these numbers
   a 503   **b** 6 192   **c** 4 018   **d** 102·4.

7 Find the missing number in each of the following. Write the answers only.
   a $607 = 407 + \square$ **b** $982 = 952 + \square$
   c $1\,000 + 1\,427 = \square$ **d** $300 + 1\,591 = \square$
   e $879 - 50 = \square$ **f** $1\,735 - 700 = \square$
   g $10·21 - \square = 0·21$ **h** $39·4 - \square = 30$

Tens number system

A

1 By counting small squares find what decimal fraction of a whole one or unit is each of the following

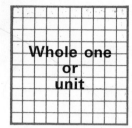

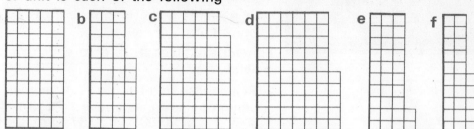

2 On graph paper draw a square like the one above to represent a whole one or unit. Shade 39 hundredths.

3 Draw similar squares and shade
   a 4 tenths   **b** 17 hundredths
   c 83 hundredths   **d** 7 hundredths.

4 Write each of the above as a decimal fraction.

B Write as decimal fractions

	a	b	c	d
1	7 tenths	19 tenths	36 tenths	70 tenths
2	23 hundredths	56 hundredths	90 hundredths	174 hundredths
3	5 hundredths	9 hundredths	2 hundredths	10 hundredths
4	$\frac{6}{10}$	$10\frac{1}{10}$	$\frac{89}{100}$	$2\frac{3}{100}$

5 a 6 tenths 8 hundredths   **b** 1 tenth 7 hundredths

C

1 How many tenths in
   a 3·7   **b** 15·6   **c** 20·4 ?

2 How many hundredths in
   a 0·76   **b** 4·67   **c** 30·05
   d 7·4   **e** 10·6   **f** 50 ?

3 1 metre = 100 centimetres

Put in the decimal point and write the following as metres
   a 125 cm   **b** 308 cm   **c** 84 cm   **d** 10 cm
   e 58 cm   **f** 9 cm   **g** 430 cm   **h** 509 cm.

D

1 Multiply by 10
   a 7·2   **b** 0·46   **c** 3·04.

2 Multiply by 100
   a 0·65   **b** 5·7   **c** 9·32.

3 Divide by 10
   a 4·7   **b** 189   **c** 0·9.

4 Divide by 100
   a 3 600   **b** 670   **c** 94.

5 Write the answers only
   a 93·71 × 100 = ☐   **b** ☐ × 100 = 346
   c ☐ ÷ 10 = 8·43   **d** 247 ÷ ☐ = 2·47.

6 How many times is the figure marked A larger or smaller than the figure marked B?
   a 7$\overset{B}{7}$$\overset{A}{7}$7   **b** 9$\overset{A}{2}$·$\overset{B}{9}$3   **c** 3·8$\overset{B}{3}$$\overset{A}{}$

Tens number system money

$$\begin{array}{c} 10 \text{ TENS} \\ 100 \text{ p} \end{array} = £1{\cdot}00$$

1 TEN is 1 **tenth** of £1·00 = £0·1 which is written as £0·10.
1p is 1 **hundredth** of £1·00 = £0·01.

1 Find in new pence the value of
  a 3 tenths of £1·00   b 7 tenths of £1·00
  c 9 tenths of £1·00   d 9 hundredths of £1·00
  e 35 hundredths of £1·00
  f 94 hundredths of £1·00.

2 Write each of the following as a decimal of £1·00.
  a 2 TENS   b 5 TENS   c 9 TENS   d 17 TENS
  e 3p   f 8p   g 29p   h 75p

Fill in the missing numbers.

3 a £3·85 = £☐ ☐ TENS ☐p
  b £10·07 = £☐ ☐ TENS ☐p
  c £25·64 = £☐ ☐ TENS ☐p

4 a £1·76 = ☐ TENS ☐p
  b £9·03 = ☐ TENS ☐p
  c £14·87 = ☐ TENS ☐p

5 In the same way read each of these sums of
money to your partner who will check from the
answer book.
  a £2·86   b £0·73   c £20·50   d £0·85
  e £4·27$\frac{1}{2}$ f £0·09   g £150·25   h £327·00

6 Draw this table in your book.
Then write in the sums of money **5a** to **5h**.
The first has been done for you.

	H £100	T £10	U £1	t £0·10	h £0·01
a			2	8	6

7 Write the value of figures underlined in each of
these sums of money.
  a £5<u>3</u>0·<u>8</u>0   b £<u>9</u>·6<u>7</u>   c £2<u>43</u>·1<u>4</u>

8 Make up each of these sums of money using the
fewest possible numbers of notes and coins.
  a £1·35   b £0·24   c £0·86
  d £0·09$\frac{1}{2}$ e £5·50   f £10·75

9 Multiply by 10
  a £1·86   b £0·65   c £0·04.

10 Multiply by 100
  a £0·73   b £0·08   c £1·31.

11 Divide by 10
  a £1·60   b £9·30   c £25·40.

12 Divide by 100
  a £2·00   b £27·00   c £117·00.

Reading and writing numbers

Up to now you have been using whole numbers to **thousands**. You will often see numbers much greater than thousands in newspapers and books. Collect other examples like those given below.

50 370 spectators watched Arsenal play Chelsea

ROAD CASUALTIES in BRITAIN 393 937

CITY TO SPEND £1½ million on schools this year

If you are interested in the journeys of astronauts into space you will read about distances of millions of kilometres.

A

Million	Hundred Thousand	Ten Thousand	Thousand	Hundred	Ten	Unit
1 000 000	100 000	10 000	1 000	100	10	1

A thousand thousands is 1 million

1 Read these numbers to your partner who will check them from the answer book.
   a 30 000   b 16 580   c 113 020
   d 301 904   e 460 705   f 1 270 010

2 Write each of these numbers in figures.
   a Seventeen thousand
   b Twenty thousand and fifteen
   c Forty-two thousand six hundred and ninety-eight
   d Seventy-six thousand and thirty-nine
   e A hundred and ten thousand and eight
   f Four hundred thousand.

3 a Write 1 million in figures.
   b Write in figures
   ½ million   ¼ million   ¾ million   2¼ million.

B 1 The following are recent population estimates.
   a Liverpool is 722 010   b Cardiff is 260 170
   c Glasgow is 1 000 857   d Belfast is 406 857
   Read them to your partner.

2 Which of these towns has
   a the greatest population   b the smallest population?

3 What is the difference in population between
   a Liverpool and Glasgow   b Cardiff and Belfast?

C 1

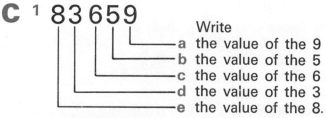

83 659
   Write
   a the value of the 9
   b the value of the 5
   c the value of the 6
   d the value of the 3
   e the value of the 8.

2 In the same way write the value of each figure in these numbers
   a 41 903   b 508 170.

3 Write the value of the figures underlined.
   a 30 260   b 100 607
   c 370 291   d 2 407 509

Round numbers

approximations

A

Arrange each group of numbers to make
a the largest possible number
b the smallest possible number.

1 5   7   2   9
2 2   7   4   1   9
3 8   6   0   5   7   2

4 How many whole **hundreds** are there in each of these numbers?
  a 709   b 4 023   c 70 180

5 How many whole **thousands** are there in each of these numbers?
  a 8 070   b 39 239   c 207 306

Large numbers are often 'rounded off' when approximate answers only are required. This means that the numbers are stated 'to the nearest hundred' or 'to the nearest thousand' etc.

B

Round numbers to the nearest hundred

| 600 | 610 | 620 | 630 | 640 | 650 | 660 | 670 | 680 | 690 | 700 |

1 Four of the numbers above are nearer to 600 than 700.
  Which are they?

2 Four of the numbers are nearer to 700 than 600.
  Which are they?

  650 is half-way between 600 and 700.

  The half-way number is always taken to the round number above, in this case 700.

3 Write each of these numbers to the nearest hundred
  a 620   b 670   c 650   d 613   e 648   f 687.

4 Round off these numbers to the nearest hundred
  a 873   b 987   c 1 150   d 1 329   e 2 310
  f 5 550   g 6 838   h 10 111   i 9 613   j 14 466.

C

Round numbers to the nearest thousand

| 6 000 | 6 100 | 6 200 | 6 300 | 6 400 | 6 500 | 6 600 | 6 700 | 6 800 | 6 900 | 7 000 |

1 Which of the numbers above are nearer to 6 000 than to 7 000?

2 Which of the numbers are nearer to 7 000 than to 6 000?

  6 500 is half-way between 6 000 and 7 000.

  The half-way number is always taken to the round number above, in this case 7 000.

3 Write each of these numbers to the nearest thousand
  a 6 311   b 6 500   c 6 850   d 9 762   e 10 291
  f 12 800   g 14 500   h 13 017   i 16 960   j 19 703.

D

Visitors to the Zoo

Week ending	Number of visitors
4th Aug	12 192
11th Aug	30 972
18th Aug	15 036
25th Aug	27 751
1st Sept	29 625

1 During which week was the attendance at the Zoo
  a the largest   b the smallest?

2 Find the total number of people who attended the Zoo during the period.

3 Write the table in your book giving the numbers to the nearest hundred.

4 From your table find
  a the approximate total attendance
  b the difference between the actual total and the approximate total.

5 Repeat questions 3 and 4 'rounding off' the numbers to the nearest thousand.

5

Round numbers

approximations

A

1 Which of these fractions are
equal to $\frac{1}{2}$ greater than $\frac{1}{2}$ less than $\frac{1}{2}$?
Use the signs $=$ $<$ or $>$ in writing the answers
e.g. $\frac{4}{8} = \frac{1}{2}$ $\frac{3}{4} > \frac{1}{2}$ $\frac{1}{4} < \frac{1}{2}$.
a $\frac{7}{10}$ b $\frac{1}{3}$ c $\frac{5}{6}$ d $\frac{3}{8}$ e $\frac{2}{6}$ f $\frac{5}{10}$

2 Write each of the following to the nearest whole one.
a $1\frac{7}{10}$ b $5\frac{1}{3}$ c $3\frac{5}{6}$ d $6\frac{3}{8}$ e $9\frac{2}{5}$ f $2\frac{5}{10}$

3 Write each of the following to the nearest whole one.
a $13\frac{1}{4}$ b $27\frac{9}{10}$ c $9\frac{1}{2}$ d $39\frac{3}{4}$ e $20\frac{5}{8}$ f $48\frac{3}{5}$
Now write each of the above to the nearest ten.

4 Write each of the following to the nearest whole one.
a 0·8   b 7·2   c 12·6   d 19·5   e 40·1   f 49·7

5 Write to the nearest £
a £11·96   b £15·20   c £21·50   d £9·08   e £17·49   f £0·76.

B

By 'rounding off' numbers you can easily obtain approximate
answers to many calculations.
The approximate answers provide a useful check to your working.
Look at these examples and complete columns **a** and **b**.

	Example	Round numbers	a Approximate answer	b Actual answer
1	713+871+537	700+900+500		
2	921 − 489	900 − 500		
3	706 × 9	700 × 9		
4	392 ÷ 4	400 ÷ 4		
5	£5·83 × 8	£6·00 × 8		
6	£14·75 ÷ 5	£15·00 ÷ 5		

7 Draw a table like the one above, and complete it using the following
examples.
a 9·8 × 4   b 27·2 × 8   c 39·5 × 6   d 99·7 × 9
e 51·6 ÷ 4   f 134·5 ÷ 5   g 161·35 ÷ 7   h 99·2 ÷ 8

8 Now use the sign $<$ or $>$ to show whether the approximate answer
is less than or greater than the actual answer.

C

Three answers are given to each of these examples.
By finding an approximate answer you can see that two are wrong.
Write the correct answer and check it by working the example.

1 264 × 4   (270   1 056   8 254)
2 872 ÷ 8   (19   203   109)
3 $38\frac{1}{2}$p × 6   (£2·31   £23·10   £38·50)
4 £8·73 ÷ 9   (£1·27   £9·70   97p)

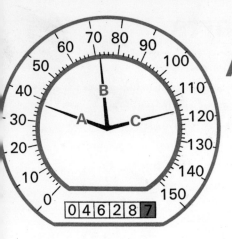

Dials and scales

A There are many instruments which show measurements in different kinds of units by means of a scale. Here are some common examples. You can find many others for yourself.

1 This picture shows a speedometer which is fitted in a car. It measures speed in kilometres per hour (km/h).

   a What is the maximum speed which can be shown on this speedometer?

   b What does a small division on the scale represent?

   c What is the speed of the car when the pointer is at A, at B, at C?

   d This speedometer also shows in km the total distance the car has travelled. How many more km must it travel before it shows 5 000 km?

2 This picture shows the dial of a weighing machine.

   a What is the maximum weight for which it can be used?

   b What weight does a small division represent?

   c Write the weight when the pointer is at A, at B, at C.

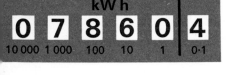

3 Most homes have electricity to give light and heat. This electricity has to be paid for, so the amount which is used is measured in units (kW h) by means of a meter.

   a Read the meter shown in the picture.

   b What does kW h stand for?

   c Find the meter reading after 60 more units have been used.

B

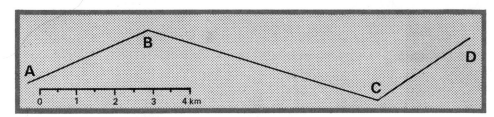

1 Use the scale to find the actual distances from

   a A to B   b B to C   c C to D.

   d Find the actual distance from A to D 'as the crow flies'.

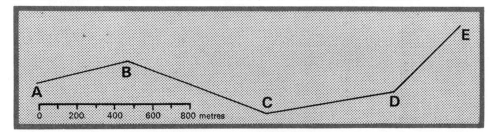

2 Use the scale to find the actual distances from

   a A to B   b B to C   c C to D   d D to E.

   e Find the actual distance from A to E 'as the crow flies'.

7

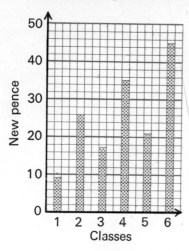

Kingsbridge School

Graphs and charts

A

The column graph shows how much money was collected by each class in Kingsbridge School for World Children's Day.

1 Name the units shown on the scale on the vertical axis.
2 How much does a small division on this scale represent?
3 What does the scale on the horizontal axis tell you?
4 a How much money did each class collect?
  b Find the total amount collected.
5 a What is the average?
  b Which classes collected more than the average, which less than the average?

B

Abbey School   World Children's Day Fund

Class	1	2	3	4	5	6	7	8	9	10
Cash	40p	60p	37p	56p	52p	46p	42p	51p	33p	43p

The table tells you how much was collected by each class in the Abbey School.

1 On squared paper draw the horizontal and vertical axes and mark the scales. This time you can begin the vertical scale at 30p. Why?
2 Draw a column graph to show how much each class collected.
3 Find
  a the total amount for the school
  b the average per class.
4 a Draw a line across the graph to show the average.
  b Which classes collected more than the average, which less than the average?

C

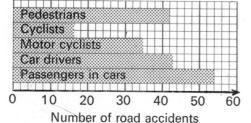

Number of road accidents

The graph gives the number of road users who were in road accidents in a town during the year.

1 Find the number for each group of road users.
2 What is the total number of accidents?
3 Give the reason why the greatest number of people involved in road accidents is of passengers in cars.

D

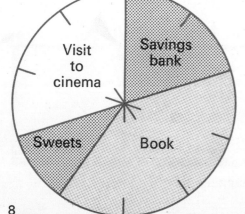

This kind of diagram is called a **pie chart.**
It shows how Katherine spent a present of £1·00.

1 A pie chart is always drawn as a circle.
Into how many equal parts has it been divided?
2 a Write as a decimal fraction the part of her money she spent on each item.
  b How much is this on each item?
3 Complete
Katherine saved _____ as much as she spent on sweets.
She spent _____ times as much on the book as she spent on sweets.
4 Draw a pie chart to show how John spent
$\frac{3}{8}$ of his money on lunch, $\frac{1}{4}$ on bus fares,
$\frac{1}{8}$ on ice cream and the rest on sweets.
What fraction is spent on sweets?

8

A
1	6 + 3
2	5 + 5
3	7 + 6
4	4 + 8
5	5 + 9
6	8 + 6
7	3 + 9
8	7 + 7
9	5 + 8
10	6 + 6
11	9 + 7
12	7 + 5
13	4 + 9
14	9 + 6
15	8 + 8
16	7 + 8
17	4 + 7
18	9 + 9
19	6 + 5
20	2 + 9

Number and money
addition practice
B

Write the answers only.

	a	b	c	d
1	6 + 19	28 + 7	77 + 9	5 + 96
2	30 + 14	21 + 20	40 + 48	93 + 10
3	23 + 17	42 + 28	15 + 35	36 + 64
4	13 + 29	66 + 17	28 + 55	36 + 38

5	6	7
89	59·8	3 204
403	164·7	987
215	38·6	1 435

8 £	9 £	10 £
2·38	1·97	10·42
6·15	16·08	75·74
4·07	5·85	134·91

Write the answers as £'s. Add across if you can.

11 37p + 44p + 38p   12 52p + 75p + 86p

13 $24\frac{1}{2}$p + 47p + $32\frac{1}{2}$p   14 $62\frac{1}{2}$p + $76\frac{1}{2}$p + $92\frac{1}{2}$p

Write the following in columns and then add.

15	1 605	307	2 416
16	127	419	6 211
17	£8·65	£9·43	£0·85
18	£0·$17\frac{1}{2}$	£0·03	£4·98
19	£18·63	$27\frac{1}{2}$p	$52\frac{1}{2}$p	£6·30
20	£3·80	$5\frac{1}{2}$p	£14·39	£0·$86\frac{1}{2}$

C

1 Increase **a** 940 by 209 **b** £4·85 by 38p.

2 How many days are there from 1st March to 31st August? The dates are inclusive.

3 The sides of a field measure 193 m, 207 m, 348 m and 287 m.
 a Name the shape of the field. **b** Find its perimeter.

4 Mother spent these sums of money in different shops, £2·84, $37\frac{1}{2}$p, £1·06 and £2·00. Her bus fares cost 20p. How much did she spend altogether?

5 The total distance a car has travelled is given on the odometer which shows 5 983·6 km.
 Find what it will register after a journey of 416·5 km.

6 Find the total value of the notes and coins shown in each row of the table below.
 (You will have five answers.)

7 Mark these answers and correct any mistakes. Then find the total sum of money.

	Notes		Coins					
	£5	£1	50p	10p	5p	2p	1p	$\frac{1}{2}$p
a		1		3	1		1	2
b			1	2	3	5		
c	1	1	2		4	3	6	1
d	1		1	8	4		3	1
e	1	1	4	3		4	2	4

9

subtraction practice

A

1	9 — 4
2	10 — 6
3	11 — 5
4	13 — 9
5	12 — 7
6	15 — 6
7	14 — 8
8	16 — 7
9	11 — 2
10	12 — 9
11	14 — 7
12	17 — 8
13	15 — 7
14	11 — 4
15	13 — 7
16	16 — 8
17	14 — 5
18	12 — 8
19	18 — 9
20	13 — 5

B

Write the answers only.

	a	b	c	d
1	23 — 4	37 — 9	51 — 7	95 — 8
2	47 — 7	28 — 20	73 — 40	87 — 30
3	40 — 9	70 — 6	50 — 8	30 — 2
4	30 — 17	60 — 24	40 — 33	70 — 51
5	25 — 17	58 — 49	43 — 27	72 — 18

6	228	7	76·3	8	5 274
	— 99		— 30·5		— 3 196

	£		£		£
9	3·70	10	4·17	11	16·07
	— 1·63		— 3·26		— 9·98

	£		£		£
12	2·43	13	5·82	14	17·43
	— 0·76		— 0·07		— 9·38

Check the answers to examples **6-14** by addition.

Set down the following and then subtract.

15 6 000 — 5 715
16 1 317 — 939
17 7 500 — 2 750
18 3 410 — 1 716
19 86½p — 39p
20 58p — 27½p
21 £1·03 — 85p
22 £3·60 — 93½p
23 £8·16 — £6·09½
24 £19·31 — £12·28

C

1 Decrease a 1 010 by 246   b £4·50 by 64½p.
2 Find the change
   a from a £1 note after paying for a pair of socks costing 32½p
   b from a £5 note after paying for a pair of shoes costing £2·45
   c from a £10 note after paying for a coat costing £8·25.
3 Jane saved 6 TENS and 14 FIVES. How much more must she save to buy a tennis racket costing £3·50?
4 a How many more pupils are there at the Junior School than at the High School?
   b There are 328 boys in the Junior School. How many girls are there?
   c There are 289 girls in the High School. How many boys are there?
   d How many more boys are there in the Junior School than in the High School?
   e How many fewer girls are there in the High School than in the Junior School?

Grange
Junior School
Number on Roll
687

Heaton
High School
Number on Roll
571

Time
hours and minutes

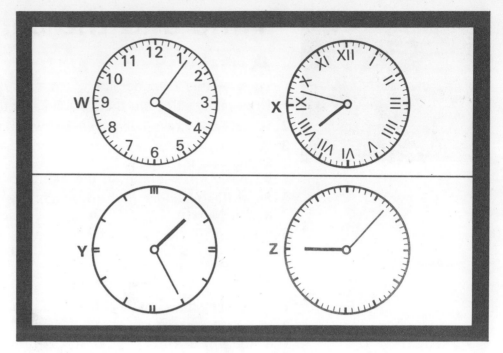

A The pictures show different kinds of watch and clock faces.
Clocks W and X show morning times, clocks Y and Z afternoon times.

1 Write in words and figures the time shown on each clock.

2 Find the correct time if
   a Clock W is $\frac{1}{4}$ h fast   b Clock X is 18 min slow
   c Clock Y is 6 min slow   d Clock Z is 12 min fast.

At the Royal Observatory in Sussex, astronomers make observations from the stars on which they base the **standard** of time. This standard is called Greenwich Mean Time (G.M.T.).

3 British Standard Time (B.S.T.) is 1 hour in front of Greenwich Mean Time. The clocks above show B.S.T. Write G.M.T. in each case.

B 1 Find the number of minutes to the next complete hour from
   a 10.25   b 9.47   c 6.18   d 2.03   e 1.32.

2 How many hours and minutes to **midday** or **noon** from
   a 11.45 a.m. b 10.20 a.m. c 9.05 a.m. d 10.53 a.m. e 8.17 a.m.?

3 How many hours and minutes to **midnight** from
   a 11.10 p.m. b 10.40 p.m. c 9.25 p.m. d 10.47 p.m. e 8.13 p.m.?

4 Find how many hours and minutes there are between
   a 9.20 a.m. and 11.10 a.m.   b 4.35 p.m. and 8·05 p.m.
   c 8.33 a.m. and 10.09 a.m.   d 1.27 p.m. and 3.16 p.m.
   e 11.10 a.m. and 12.50 p.m.   f 10.20 p.m. and 2.35 a.m.
   g 9.53 a.m. and 1.10 p.m.   h 9.16 p.m. and 2.10 a.m.

5 To bake a fruit cake takes 2 h 20 min. If Mother puts the cake into the oven at 3.55 p.m. at what time must she take it out?

6 To roast a turkey takes 20 min for each $\frac{1}{2}$ kg plus 20 min.
   a How long will it take to roast a turkey weighing $4\frac{1}{2}$ kg?
   b At what time must Mother put it into the oven if she wishes it to be cooked by 1 p.m.?

BUS SERVICE NO.	Times from this stop
21	To the CITY CENTRE 20 min to and past each hour
	To the TERMINUS 15 min to and past each hour

C

The buses on Service No. 21 travel
from the **City Centre** to the **Terminus**
from the **Terminus** to the **City Centre**.

1 What is the time of the next bus from this stop to the centre after a 10 a.m. b 3.45 p.m.?

2 How long must I wait for a bus to the terminus if I arrive at the stop at a 8.25 a.m. b 8.50 p.m.?

11

FEBRUARY				
Sun	1	8	15	22
Mon	2	9	16	23
Tues	3	10	17	24
Wed	4	11	18	25
Thurs	5	12	19	26
Fri	6	13	20	27
Sat	7	14	21	28

Time and the calendar

A

Look at this calendar for the month of February.

1 How many days are there in the month?
2 Is this February in a leap year? How do you know?
3 Write the date of
  a the second Friday
  b the fourth Monday in the month.
4 From this calendar find on which day was
  a 6th March   b 24th January.
5 Which of these years are leap years
  1840   1970   1976   1990   2004?

B

1 How many days are there in a 1 year b 1 leap year?
2 Write the names of
  a the 4th month   b the 7th month
  c the 10th month of the year.
  How many days are there in each of these months?
3 Write the dates of
  a all the Thursdays in July if the 1st of the month
    is a Tuesday
  b all the Fridays in November if the 1st of the
    month is a Monday.
4 In a particular year 1st December was on
  Thursday. On which day was
  a Christmas Day b New Year's Day of the next year?
5 Find the number of days not counting the first day
  a from 3rd March to 21st March
  b from 13th March to 4th April
  c from 19th July to 10th Aug
  d from 25th Oct to 7th Dec.
6 These dates are inclusive. Find the number of days
  a from 18th Jan to 16th Feb
  b from 3rd April to 12th May
  c from 7th June to 29th July
  d from 10th Sept to 5th Nov.

C

Name	Date of birth		
Philip	21	1	61
Jane	24	5	63
Tony	18	9	59
Susan	9	10	62

1 Write in full the birthdays of each of these children.
2 Which child is a the oldest b the youngest?
3 Find their ages in years and months on 31st Dec
  1980. Count a half month or more as a whole month.
4 By how many months is
  a Jane older or younger than Philip
  b Tony older or younger than Susan?

D

You are now living in the 20th century.

1 a What will be the date when you are 50 years old?
  b In which century will you then be living?
2 In which centuries did these Kings and Queens
  reign?
  a William the Conqueror (1066–1087)
  b Elizabeth I (1538–1603)
  c Alfred the Great (871–901)
  d Victoria (1837–1901).

A

	a	b
1	3 × 9	5 × 8
2	4 × 6	9 × 5
3	7 × 9	7 × 6
4	6 × 6	4 × 9
5	6 × 8	9 × 3
6	7 × 3	4 × 5
7	8 × 5	6 × 9
8	4 × 7	5 × 5
9	5 × 4	8 × 3
10	9 × 6	7 × 8
11	5 × 7	9 × 7
12	8 × 4	6 × 5
13	3 × 8	7 × 4
14	5 × 9	8 × 8
15	6 × 7	4 × 8
16	5 × 6	9 × 9
17	9 × 8	8 × 6
18	8 × 7	7 × 5
19	9 × 4	6 × 4
20	7 × 7	8 × 9

Number and money
multiplication practice

B

Write the answers only.

	a	b	c
1	9p × 7	16p × 5	13p × 6
2	8p × 10	12p × 8	23p × 4
3	$3\frac{1}{2}$p × 9	$11\frac{1}{2}$p × 7	$29\frac{1}{2}$p × 3

Mark the answers and correct any mistakes.
Then write the answers as £'s.

C

Write the answers only as £'s.

	a	b	c
1	37p × 4	29p × 6	41p × 8
2	19p × 9	35p × 5	28p × 7
3	$16\frac{1}{2}$p × 10	$76\frac{1}{2}$p × 4	$47\frac{1}{2}$p × 5

D

Write the answers only when you can.

	a	b	c
1	87 × 6	208 × 9	354 × 7
2	290 × 5	607 × 4	863 × 8
3	15·3 × 8	47·2 × 6	70·6 × 7
4	86·5 × 3	29·4 × 5	39·8 × 9
5	£0·06 × 7	£0·09$\frac{1}{2}$ × 4	£0·23 × 8
6	£1·34 × 5	£3·55 × 6	£10·87 × 3

7 Multiply by 10

   a 590   b 63·4   c £0·87   d £7·32.

E

1 a Find the product of 65 and 7.
  b $x \div 9 = 29$. Find the value of x.
  c 93 × 6 = 558
    Write the answer only to 93 × 12, to 93 × 18.

2 A turkey weighing $4\frac{1}{2}$ kg costs $34\frac{1}{2}$p per $\frac{1}{2}$ kg.
  Find the price of the turkey.

3 In a school there are 7 classes.
  The average number of children per class is 35.
  Find the number of children in the school.

4 A bicycle can be bought for six monthly payments
  each of £4·75. What is the cost of the bicycle?

5 How far can a car travel in $5\frac{1}{2}$ h at an average
  speed of   a 56 km/h   b 76 km/h?

6 The adult railway fare to a scout camp is £2·52.
  Find the total cost of the fares for
  a 7 boy scouts who all travel at half price
  b 2 scoutmasters and 10 boy scouts.

Square and cubic numbers

A Square numbers

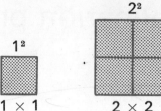
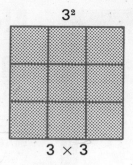

1^2 2^2 3^2

1×1 2×2 3×3

1 Write the value of 1^2, 2^2, 3^2.

2 On squared paper draw a diagram to show 5^2, 7^2. Write the square number which is equal to each.

3 Write the square number which is equal to
   a 6^2 b 8^2.

4 Write the first ten square numbers.

5
$$1^2 = 1$$
$$2^2 = 4 = 1 + 3$$
$$3^2 = 9 = 1 + 3 + \square$$
$$4^2 = 16 = 1 + 3 + 5 + \square$$
$$5^2 = 25 = 1 + 3 + 5 + 7 + \square$$

a Copy this table and fill in the missing numbers.
b Continue the table for square numbers to 8^2. Notice that each square number equals the **sum of consecutive odd numbers.**

6 From what you have discovered find, in the shortest way, the sum of the first ten odd numbers.

B Cubic numbers

1

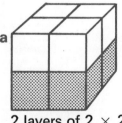
a

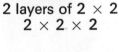

2 layers of 2×2
$2 \times 2 \times 2$

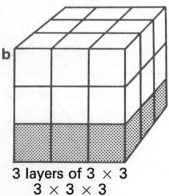
b

3 layers of 3×3
$3 \times 3 \times 3$

Get some cubes all of the same size. Build them into larger cubes as shown in diagrams **a** and **b**.

A short way of writing $2 \times 2 \times 2$ is 2^3 which is read as '2 cubed'.

1 Find the cubic number which is equal to
   a $2^3 = 2 \times 2 \times 2$ b $3^3 = 3 \times 3 \times 3$.

2 What is meant by 1^3? Find the value of 1^3.

3 Make a block of cubes to show 4^3. Then write the number equal to 4^3.

4 The first four cubic numbers are   1   8   27   64. Find the next two cubic numbers.

5 What is the number equal to 8^3?

6 Write $10 \times 10 \times 10$ as a cubic number.

7

$1^3 = 1$	$2^3 = 8$	$3^3 = 27$	$4^3 = 64$
1	$3+5$	$7+9+\square$	$13+15+17+\square$

a Copy this table and fill in the missing numbers on the bottom row.
b Mark the answers, then read the numbers on the bottom row.
   1   3   5   7   —   —   —   —   —   —.
   Notice they are the consecutive **odd** numbers.
c Find the odd numbers which when added together make 5^3 6^3.
   Check your answers by multiplying.

14

A

	a	b
1	42÷7	27÷3
2	30÷6	35÷7
3	54÷9	42÷6
4	21÷3	32÷4
5	45÷9	72÷9
6	70÷10	100÷10
7	15÷3	40÷8
8	64÷8	36÷6
9	36÷9	81÷9
10	49÷7	36÷4
11	24÷8	28÷7
12	20÷5	72÷8
13	56÷7	24÷3
14	28÷4	45÷5
15	63÷9	63÷7
16	32÷8	25÷5
17	40÷5	56÷8
18	24÷4	48÷6
19	27÷9	18÷3
20	48÷8	54÷6

Number and money
division practice

B

Write the answers only. Work to $\frac{1}{2}$p when necessary.

	a	b	c
1	81p ÷ 9	35p ÷ 7	56p ÷ 8
2	69p ÷ 3	75p ÷ 5	84p ÷ 6
3	78p ÷ 4	85p ÷ 10	92p ÷ 8
4	$52\frac{1}{2}$p ÷ 5	$94\frac{1}{2}$p ÷ 9	$87\frac{1}{2}$p ÷ 7

Mark the answers and correct any mistakes.
Then write the answers as £'s.

C

Write the answers only as £'s.
Work to $\frac{1}{2}$p when necessary.

	a	b	c
1	3)£1·44	6)£3·42	4)£2·96
2	10)£7·80	8)£5·36	5)£4·35
3	7)£1·64$\frac{1}{2}$	9)£3·82$\frac{1}{2}$	4)£3·50
4	8)£5·80	5)£4·77$\frac{1}{2}$	6)£5·25

Mark the answers and correct any mistakes.
Then write the answers as new pence only.

D

Write the answers only when you can. Some examples have remainders.

	a	b	c	d
1	702 ÷ 2	953 ÷ 9	1 315 ÷ 5	3 900 ÷ 7
2	1 209 ÷ 3	2 825 ÷ 6	3 632 ÷ 4	4 500 ÷ 8
3	4)193·2	7)426·3	2)141·8	8)859·2
4	3)291·9	6)738·6	9)596·7	5)318·0
5	4)£27·12	8)£9·20	9)£72·54	3)£15·21
6	7)£22·26	5)£30·95	2)£16·14	6)£58·56

E

1 a Find the remainder when 2 917 is divided by 9.
  b Write the following as mixed numbers
     $\frac{79}{5}$ $\frac{110}{8}$ $\frac{115}{10}$.
  c Find the missing number □ × 7 = 117·6.

2 Share a prize of £5·00 equally among 7 children.
   How much does each child receive and how much
   remains?

3 Divide by 10
   a 756   b 89·4   c £19·30.

4 A joint of meat weighing $2\frac{1}{2}$ kg costs £1·50.
   Find the price of the meat
   a per $\frac{1}{2}$ kg   b per kg   c per 200 g.

5 Find the average of these numbers
   102·6   98·4   87·3   74·1.

Fractions

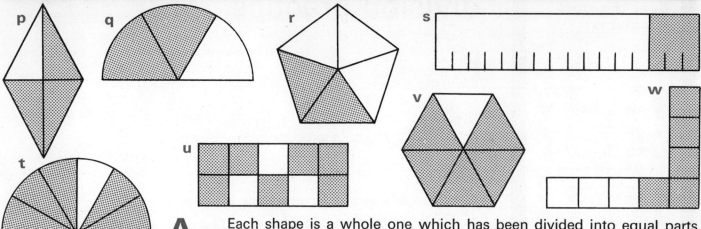

A Each shape is a whole one which has been divided into equal parts.

1 What fraction of each whole one is **a** shaded **b** unshaded?

2 Write a sentence telling how each fraction is found.
The first example is done for you
p To find $\frac{3}{4}$, divide the whole one into 4 equal parts, take 3 of them.

REMEMBER
The number below the line (the DENOMINATOR) tells you into how many equal parts the whole one is divided. The number above the line (the NUMERATOR) tells you how many of the parts have been taken.

B 1 In the fraction $\frac{7}{9}$
  a What is the 9 called? What does it tell you?
  b What is the 7 called? What does it tell you?

2 **a** Write the fraction with 4 as the numerator and 7 as the denominator.
  b Write the fraction in words.

3 A birthday cake is cut into 10 equal slices, 9 of which are eaten. What fraction of the cake remains?

4 Of 15 children who travel to school on a bus, 4 are girls.
What fraction of the children are **a** girls **b** boys?

5 In a class of 30 children, 7 are absent.
What fraction of the class is **a** absent **b** present?

6 A sum of money is divided into equal shares.
Susan has 2 shares, Molly 3 shares and Joan 4 shares.
What fraction of the money does each girl receive?

7 John scored 17 marks out of 20 and Mary 13 out of 20.
What fraction of the marks did each score?

8 Write as fractions
  a 23 out of 50   **b** 9 out of 100.

C 1 Write each of the following as an improper fraction, then as a mixed number.
  a 7 quarters **b** 8 thirds **c** 13 tenths **d** 12 fifths

2 Change these to improper fractions
  a $1\frac{2}{3}$ **b** $1\frac{3}{5}$ **c** $2\frac{3}{4}$ **d** $2\frac{3}{8}$.

3 Change these to mixed numbers
  a $\frac{17}{4}$ **b** $\frac{22}{3}$ **c** $\frac{43}{10}$ **d** $\frac{39}{5}$.

16

Fractions

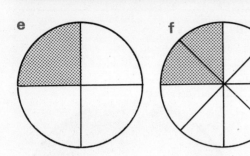

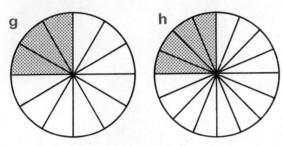

A 1 Into how many equal parts has each whole one
   e,   f,   g,   h been divided?

2 Write the following putting in the missing numerator or denominator.

a $\frac{1}{4} = \frac{\square}{8} = \frac{3}{\square} = \frac{\square}{16}$

b $\frac{3}{4} = \frac{6}{\square} = \frac{\square}{12} = \frac{12}{\square}$

c Write another fraction which is equal to $\frac{1}{4}$ and another equal to $\frac{3}{4}$.

3 In the same way write and complete the following.

a $\frac{1}{3} = \frac{\square}{6} = \frac{3}{\square} = \frac{\square}{12}$ b $\frac{2}{3} = \frac{4}{\square} = \frac{\square}{9} = \frac{8}{\square}$

c Write another fraction which is equal to $\frac{1}{3}$ and another equal to $\frac{2}{3}$.

Look again at the answers to **2** and **3** above. Notice that when the **numerator** and **denominator** of a fraction are **multiplied or divided** by the **same number** the value of the fraction is **not changed.**

B   Write the following fractions putting in the missing numerator or denominator for *x* or *y*.

1 $\frac{1}{2} = \frac{x}{16}$ 2 $\frac{7}{8} = \frac{14}{y}$ 3 $\frac{3}{10} = \frac{9}{y}$ 4 $\frac{4}{5} = \frac{12}{y}$

5 $\frac{7}{10} = \frac{x}{20}$ 6 $\frac{2}{3} = \frac{x}{15}$ 7 $\frac{5}{6} = \frac{20}{y}$ 8 $\frac{2}{5} = \frac{x}{20}$

9 $\frac{1}{10} = \frac{x}{100}$ 10 $\frac{7}{20} = \frac{x}{100}$ 11 $\frac{3}{50} = \frac{x}{100}$ 12 $\frac{1}{5} = \frac{x}{100}$

C   Divide both numerator and denominator of each of these fractions by the same number. This is called **cancelling** the fraction.

1 a $\frac{6}{10}$ b $\frac{9}{12}$ c $\frac{15}{20}$ d $\frac{12}{15}$ e $\frac{10}{16}$

By cancelling reduce each of these fractions **to its lowest terms** (fractions with the smallest possible numbers in the numerator and denominator).

2 a $\frac{6}{8}$ b $\frac{4}{16}$ c $\frac{8}{12}$ d $\frac{4}{20}$ e $\frac{6}{18}$

3 a $\frac{10}{20}$ b $\frac{9}{15}$ c $\frac{8}{10}$ d $\frac{6}{24}$ e $\frac{10}{15}$

4 a $\frac{50}{100}$ b $\frac{10}{100}$ c $\frac{30}{100}$ d $\frac{5}{100}$ e $\frac{25}{100}$

D   Cancelling can sometimes be used instead of division to find an answer more easily.

e.g. $84 \div 16 = \frac{84}{16} = \frac{84 \div 4}{16 \div 4} = \frac{21}{4} = 5\frac{1}{4}$.

By cancelling find the answer to the following.

1 a $90 \div 15$ b $66 \div 12$ c $72 \div 16$ d $125 \div 25$

2 A note book costs 14p.
   How many can be bought for 84p?

3 How many kg of apples costing 18p per kg can be bought for 99p?

17

Fractions

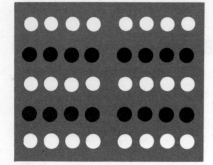

A

1 **a** Find the total number of beads in the box.
   b How many beads are white?
      How many are black?
   c Write in the lowest terms the fraction of the beads which are white, the fraction which are black.

2 The pie chart shows how Peter spent a whole day.
   a Into how many equal parts is the chart divided? Why?
   b What period of time does each part represent?
   c What fraction of the day does Peter spend doing each of the things? Write the fractions in their lowest terms.

(Pie chart labelled: Sleeping, School, Homework, Play, Meals and TV)

3 Mary worked a number test of 50 items. She had 35 right out of the 50. What fraction in its lowest terms had she
   a right   **b** wrong?

4 20 out of 32 children in a class were girls. What fraction of the class were **a** girls **b** boys?

5 Write each of the following as a fraction in its lowest terms.
   a 12 out of 18   **b** 9 out of 30   **c** 15 out of 25
   d 10 out of 100 **e** 60 out of 100 **f** 75 out of 100

B

Find the value of

	a	b	c	d
1	$\frac{1}{3}$ of 87	$\frac{1}{5}$ of 65p	$\frac{1}{8}$ of 96	$\frac{1}{6}$ of 90p
2	$\frac{3}{4}$ of 240	$\frac{5}{6}$ of 42p	$\frac{7}{10}$ of 900	$\frac{5}{8}$ of 56p
3	$\frac{3}{8}$ of £1·20	$\frac{5}{6}$ of £3·00	$\frac{2}{3}$ of £2·85	$\frac{2}{7}$ of £1·40
4	$\frac{3}{10}$ of £12·00	$\frac{4}{5}$ of £6·50	$\frac{7}{8}$ of £10·00	$\frac{4}{9}$ of £180·00

C

1 In a class there were 20 girls which was $\frac{5}{8}$ of the total number of children. Complete the following.
   a $\frac{5}{8}$ of the total number of children = □
   b $\frac{1}{8}$ of the total number of children = □
   c $\frac{8}{8}$ or the total number of children = □.

2 Susan bought a present costing 72p which was $\frac{3}{4}$ of her savings. Find the value of
   a $\frac{1}{4}$ of her savings   **b** the whole of her savings.

3 $\frac{3}{5}$ of a number is 42. Find the number.

4

Bus fares and cinema	Savings bank	Sweets

The diagram shows how a boy spent his pocket money.
   a What fraction of his pocket money did he pay out on each item?
   b He paid 28p for the bus fares and the cinema. How much did he save?
      How much did he spend on sweets?

Fractions

Addition

Work the examples in Section **A**.
Mark the answers and correct any mistakes.
Then go on to Section **B**.

Make sure that all the answers are in their lowest terms.

A

	a	b	c	d
1	$\frac{3}{8} + \frac{1}{8}$	$\frac{7}{10} + \frac{1}{10}$	$\frac{2}{5} + \frac{3}{5}$	$\frac{1}{6} + \frac{1}{6}$
2	$\frac{2}{3} + \frac{1}{3}$	$\frac{1}{4} + \frac{1}{4}$	$\frac{5}{6} + \frac{1}{6}$	$\frac{3}{10} + \frac{1}{10}$
3	$\frac{1}{8} + \frac{5}{8}$	$\frac{1}{5} + \frac{3}{5}$	$\frac{5}{12} + \frac{7}{12}$	$\frac{7}{16} + \frac{5}{16}$

The answers to the following are improper fractions.
Change them to mixed numbers.

4	$\frac{5}{8} + \frac{7}{8}$	$\frac{2}{3} + \frac{2}{3}$	$\frac{9}{10} + \frac{3}{10}$	$\frac{5}{6} + \frac{5}{6}$
5	$\frac{3}{5} + \frac{4}{5}$	$\frac{7}{10} + \frac{9}{10}$	$\frac{7}{8} + \frac{5}{8}$	$\frac{7}{12} + \frac{11}{12}$

B

Example 1

$$\frac{5}{8} + \frac{1}{4}$$
$$= \frac{5}{8} + \frac{2}{8}$$
$$= \frac{7}{8}$$

Example 2

$$\frac{2}{3} + \frac{1}{2}$$
$$= \frac{4}{6} + \frac{3}{6}$$
$$= \frac{7}{6} = 1\frac{1}{6}$$

To add unlike fractions, change them to fractions with the same name (**denominator**). Look at the examples, then work the following.

	a	b	c
1	$\frac{3}{8} + \frac{1}{2}$	$\frac{3}{10} + \frac{2}{5}$	$\frac{1}{2} + \frac{1}{10}$
2	$\frac{1}{2} + \frac{7}{8}$	$\frac{5}{6} + \frac{1}{3}$	$\frac{9}{10} + \frac{1}{2}$
3	$\frac{3}{4} + \frac{3}{8}$	$\frac{5}{12} + \frac{1}{2}$	$\frac{3}{5} + \frac{7}{10}$
4	$\frac{1}{2} + \frac{4}{5}$	$\frac{2}{3} + \frac{5}{6}$	$\frac{3}{4} + \frac{7}{8}$

Subtraction

Work the examples in Section **C**.
Mark the answers and correct any mistakes.
Then go on to Section **D**.

C

	a	b	c	d
1	$\frac{3}{4} - \frac{1}{4}$	$\frac{7}{8} - \frac{3}{8}$	$\frac{2}{3} - \frac{1}{3}$	$\frac{5}{6} - \frac{1}{6}$
2	$\frac{9}{10} - \frac{3}{10}$	$\frac{5}{8} - \frac{5}{8}$	$\frac{7}{12} - \frac{5}{12}$	$\frac{4}{5} - \frac{2}{5}$
3	$1 - \frac{7}{8}$	$1 - \frac{3}{10}$	$1 - \frac{1}{6}$	$1 - \frac{2}{5}$

D

Example

$$\frac{9}{10} - \frac{2}{5}$$
$$= \frac{9}{10} - \frac{4}{10}$$
$$= \frac{5}{10} = \frac{1}{2}$$

To subtract unlike fractions, change them to fractions with the same name (**denominator**). Look at the example then work the following.

	a	b	c
1	$\frac{2}{3} - \frac{1}{6}$	$\frac{4}{5} - \frac{1}{2}$	$\frac{7}{10} - \frac{3}{5}$
2	$\frac{5}{8} - \frac{1}{2}$	$\frac{2}{3} - \frac{1}{2}$	$\frac{4}{5} - \frac{3}{10}$
3	$\frac{7}{8} - \frac{1}{4}$	$\frac{3}{4} - \frac{5}{8}$	$\frac{3}{4} - \frac{2}{3}$

E

Write the answers only to the following.

1 $1\frac{5}{8} - \frac{7}{8}$ 2 $1\frac{1}{3} - \frac{2}{3}$ 3 $1\frac{7}{10} - \frac{9}{10}$ 4 $1\frac{3}{5} - \frac{4}{5}$

Example

$1\frac{3}{10} - \frac{2}{5} = 1\frac{3}{10} - \frac{4}{10}$
It is impossible to take $\frac{4}{10}$ from $\frac{3}{10}$ so change the whole one to tenths and add it to $\frac{3}{10}$ making $\frac{13}{10}$.
$\frac{13}{10} - \frac{4}{10} = \frac{9}{10}$.

Look at the example then work the following.

5 $1\frac{3}{8} - \frac{3}{4}$ 6 $1\frac{1}{10} - \frac{4}{5}$ 7 $1\frac{2}{5} - \frac{9}{10}$

8 $1\frac{1}{2} - \frac{7}{8}$ 9 $1\frac{2}{5} - \frac{1}{2}$ 10 $1\frac{1}{4} - \frac{2}{3}$

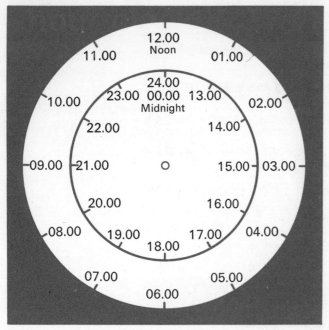

24-hour clock

A

To remind you

The hours are counted from 00.00 to 12.00 (midnight to noon), 12.00 to 24.00 (noon to midnight), a.m. and p.m. are therefore unnecessary.

When writing hours and minutes always use four figures (including 0 where necessary), the first two for **hours**, the last two for minutes. A point usually separates the hours from the minutes.

1 Write these times as 24-hour clock times.
  a 4.13 a.m.   b 7.20 p.m.   c 3.12 a.m.
  d 9.50 p.m.   e 11.56 a.m.   f 12.03 a.m.

2 Write these times as 12-hour clock times.
  a 06.25   b 15.37   c 09.54
  d 20.45   e 11.59   f 23.56

B

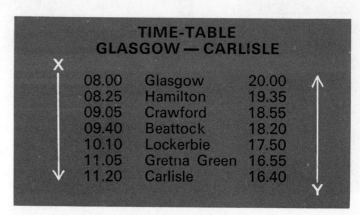

TIME-TABLE
GLASGOW — CARLISLE

X			Y
08.00	Glasgow	20.00	
08.25	Hamilton	19.35	
09.05	Crawford	18.55	
09.40	Beattock	18.20	
10.10	Lockerbie	17.50	
11.05	Gretna Green	16.55	
11.20	Carlisle	16.40	

The time-table shows the times taken by two buses.
Bus X from Glasgow to Carlisle
Bus Y from Carlisle to Glasgow
Each bus stops at the same towns on the journey.

Write in 12-hour clock times when

1 Bus X leaves Glasgow, arrives Carlisle.
2 Bus Y leaves Carlisle, arrives Hamilton.
3 Bus X leaves Lockerbie, arrives Gretna Green.
4 Bus Y leaves Beattock, arrives Crawford.
5 Bus X leaves Crawford, arrives Lockerbie.
6 Bus Y leaves Gretna Green, arrives Lockerbie.

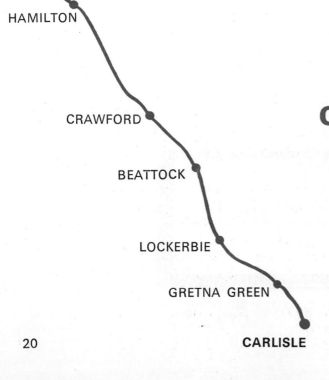

C

Copy the following table in your book.
Use the bus time-table to complete it.

	From	To	Bus times dep	arr	Time taken h min
1	Glasgow	—Beattock			
2	Beattock	—Crawford			
3	Carlisle	—Lockerbie			
4	Lockerbie	—Gretna Green			
5	Crawford	—Carlisle			
6	Gretna Green	—Hamilton			

24-hour clock

BRITISH RAIL		
LONDON King's Cross	LEEDS City	BRADFORD Exchange
11.30	14.15	
13.20	16.36	17.18
15.55	18.37	
16.05		19.21
16.20	19.56	20.36

A

The railway time-table shows some of the trains from London (King's Cross) to Leeds and Bradford.

1 Make a list of the departure and arrival times of trains for
   a Leeds only b Bradford only c Leeds and Bradford.
2 Find the time for each journey.
3 At what time must a man leave London to keep an appointment in
   a Leeds at 6 p.m.   b Bradford at 7.30 p.m.?
4 How long will the man have to wait at
   a Leeds   b Bradford?

B

The time-tables below show three different ways of travelling from London to Paris. Look at a map and follow the routes of each journey.

CAR and BOAT	
London	dep 10.15
Dover	arr 12.15
Dover	dep 13.00
Calais	arr 14.20
Calais	dep 14.40
Paris	arr 18.45

TRAIN and HOVERCRAFT	
London	dep 12.00
Folkestone	arr 13.30
Folkestone	dep 14.20
Boulogne	arr 14.55
Boulogne	dep 15.30
Paris	arr 18.20

COACH and AIRCRAFT	
London	dep 15.20
London Airport	arr 15.50
London Airport	dep 16.25
Paris Airport	arr 17.15
Paris Airport	dep 17.40
Paris	arr 18.15

1 Write in 12-hour clock times
   a the departure time by hovercraft from Folkestone
   b the arrival time at Paris Airport
   c the departure time by car from Calais
   d the arrival time at London Airport.
2 Find the total time taken for the journey from London to Paris
   a by car and boat
   b by train and hovercraft
   c by coach and aircraft.
3 Which journey is a the fastest b the slowest?
   What is the difference in time between the fastest and slowest journeys?
4 How long does it take for the sea crossing
   a from Dover to Calais by boat
   b from Folkestone to Boulogne by hovercraft?
5 Find the waiting time for passengers at
   a Dover b Calais c Folkestone d Boulogne.
6 What is the flying time from London Airport to Paris Airport?
7 How long does it take to travel by coach from
   a London to London Airport
   b Paris Airport to Paris?

21

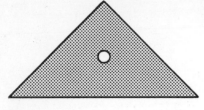

Lines and angles

A

Set squares

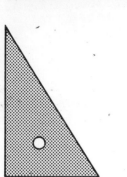

1 Get two set squares like those shown in the drawing.
  Draw round each set square and write in your drawing the measurement of each angle.

2 Use the set squares to draw these angles.
  a 45°   **b** 90° or 1 right angle   **c** 60°   **d** 30°

3 By combining the angles on the set squares draw these angles.
  a 105° (60° + 45°)   **b** 75° (45° + 30°)
  c 120°   **d** 135°   **e** 150°   **f** 165°

4 Write under each angle you have drawn either **acute angle** or **obtuse angle**.

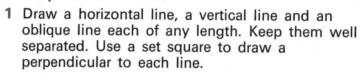

B

Perpendicular lines

1 Draw a horizontal line, a vertical line and an oblique line each of any length. Keep them well separated. Use a set square to draw a perpendicular to each line.

2 **a** Obtain a strip of paper.
     Fold it across as shown in the diagram.
  b Now fold it again so that Y rests on X.
  c Open the paper and use a ruler and pencil to mark the two creases.
  d Use a set square to measure each of the angles **m   n   o** and **p**.
  e What do you discover about the angles?
  f What can you say about the crease lines?

3 Perpendicular lines can be drawn, using a ruler and compasses only. The diagram shows how to do it.
  a Draw a line AB of any length.
  b With centre A draw an arc as shown.
  c With the same radius, from C mark off point D, and from D mark off point E.
  d Again using the same radius but with centres D and E draw two arcs intersecting (cutting each other) at F.
  e Draw a line from A through F.
  f Use your set square to show that AB and AF are perpendicular to each other.

4 **a** Draw a horizontal line 50 mm long.
  b Draw a perpendicular line at each end using a ruler and compasses.
  c Describe what you must now do to make a square, a rectangle.

5 With a ruler and compasses only draw
  a a square with sides of 65 mm
  b a rectangle 70 mm long and 45 mm wide.

Parallel and perpendicular lines

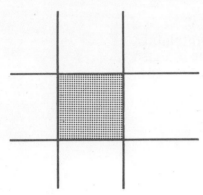

A

Parallel lines

1 Place your ruler horizontally across your paper and draw a line along each edge of the ruler. These are parallel lines.
   a How do you know?   b How far apart are they?

2 In the same way draw
   a two vertical parallel lines
   b two oblique parallel lines.

3 Use your ruler to make a drawing like the one at the top of this page. Name the shape which is shaded.

4 Use only your ruler to draw
   a a rectangle   b a rhombus   c a parallelogram.

5 a Draw a horizontal line, a vertical line and an oblique line each 60 mm long. Keep them well separated.
   b Using only a set square draw to each line three perpendiculars.
   c Find if the three perpendiculars are parallel to each other.

6 The diagram below shows a method of drawing parallel lines to a given distance apart.

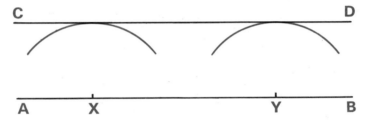

   a Draw any line AB.
   b Open the compasses to the given distance.
   c From any points, X and Y, on AB draw the arcs.
   d Draw CD so that it just touches both arcs.
   e The lines AB and CD are parallel. Can you explain why?
     Check by measuring the distances apart at any perpendicular to AB.

7 Draw parallel lines which are
   a vertical 25 mm apart   b oblique 38 mm apart
   c horizontal 43 mm apart.

B

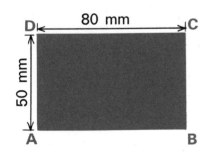

1 Using a ruler and compasses draw a rectangle of the given measurements. Letter the rectangle as shown.

2 Which of the sides are perpendicular to AB, to CD?

3 Which of the sides are perpendicular to AD, to BC?

4 Which of the sides are parallel?

5 In each of these shapes, find
   a which sides are parallel
   b which sides are perpendicular to each other.

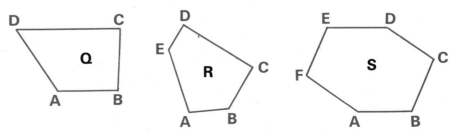

6 Write a sentence describing
   a parallel lines   b perpendicular lines.

Number Test Record
John Brown

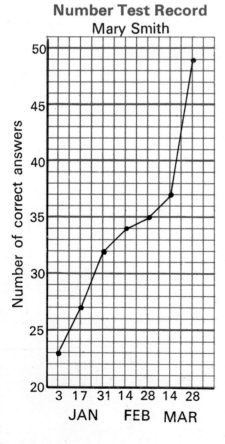

Graphs

The children in a class at Fairfield School work a number table test of 50 items each Friday fortnightly. They each keep a record of progress. Some of these are shown on these two pages.

A

This is a record of John Brown's results.

1 What is shown on the vertical axis?

2 What does one division represent on the vertical scale?

3 Why do the numbers stop at 50?

4 What is shown on the horizontal axis?

5 How many times did John take the test?

6 During which fortnight was there
   a no improvement
   b most improvement?

7 John was absent from school for several days. During which period do you think this happened? Give a reason for your answer.

8 Write the date and the result of each test.

Number Test Record
Mary Smith

B

This is the record of Mary Smith's results. Notice that she did not draw columns to show each result. To save time, she marked the height only with a dot and joined each of the dots with a thin line.

You can see her progress now quite easily.

1 There are two periods when she made rapid progress. Give the dates.

2 There is a period when her progress was slow. Give the dates.

3 Look at the slope of the line.
   a When is it steepest?
   b When is it less steep?

4 If the line remained horizontal, what would that tell you?

5 Can you describe, by looking at the two graphs, in what way Mary's progress is different from John's?

6 Write the date and result of each of Mary's tests.

24

Number Test Record
Gordon Bray

Graphs

A

This is the record of Gordon Bray's results. He made it by marking the height of each column with a dot in exactly the same way as Mary Smith.

1 Draw a table like the one below and complete it to show Gordon's score on each date.

Jan 3	Jan 17	Jan 31	Feb 14	Feb 28	Mar 14	Mar 28

2 Look at these results. By how many is the score increased each time?

3 In what way is Gordon's record different from either Mary's or John's?

4 From the table above, make Gordon's record, plotting the scores as shown.

5 Place your ruler across all the points. What do you find? Draw the line of the graph.

B

Here are the test results of two more children in the class. Look at these results carefully.

Name	Jan 3	Jan 17	Jan 31	Feb 14	Feb 28	Mar 14	Mar 28
Susan King	19	22	25	28	31	34	37
Barry Taylor	12	18	24	30	36	42	48

1 By how many does Susan's score increase each time she takes the test?

2 By how many does Barry's score increase each time he takes the test? The improvement which Susan and Barry make can be described as regular.

3 Obtain a sheet of graph paper and set out the horizontal and vertical scales as in the example. Plot Susan's scores. Then join the points by a thin line. What kind of line have you drawn?

4 On the same piece of paper, plot Barry's scores. (You will have to alter the scale first.) Join the points by a thin line. What kind of line have you drawn?

> **What you have discovered is important**
> **When quantities increase at the same rate the graph which is made will be in the form of a straight line**

5 What happens to the line of the graph if the quantities **decrease** at the same rate?

Straight line graphs

A

Straight line graphs are important because they can be used as **ready reckoners.**

Here is a simple example to show you how to make one.

Number of bars	1	2	3	4	5
Cost	5p	10p			

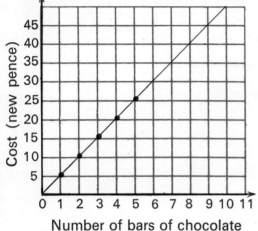

Number of bars of chocolate

1 Complete the table which shows the cost of bars of chocolate at 5p each.
Notice that the number of bars and the cost each increase at the same rate.

2 On a piece of squared paper draw the horizontal and vertical axes and mark the scale on each as shown.

3 Count 1 bar across and 5p up. Mark the point.

4 Count 2 bars across and 10p up. Mark the point.

5 In the same way plot the other pairs.

6 Join the points and you have a straight line which can be drawn as long as you wish.

7 Read from your graph the cost of
a 6 bars   b 11 bars   c 16 bars.

8 From the graph read how many bars can be bought
for   a 35p   b 50p   c 75p   d £1·00.

9 Now draw your own straight line graph to use as a **ready reckoner** for numbers of articles at 8p each.

B

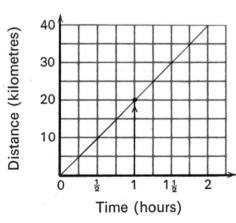

Time (hours)

Here is another straight line graph which shows a cyclist travelling at 20 kilometres per hour (km/h).

1 What does 1 division represent on the scale on
a the horizontal axis   b the vertical axis?

2 Read from the graph how far the cyclist goes in
a $\frac{1}{2}$ hour   b $1\frac{1}{2}$ hours   c 2 hours.

3 How long will it take the cyclist to travel
a 15 km   b 25 km   c 35 km?

C

1 Draw two graphs showing
a a boy walking at 6 km/h
b a train travelling at 40 km/h.

It will be necessary first to decide the scale of distances to mark on the vertical axis of each graph. Allow space to extend the scales as required.

2 Without using the graph, work out how far
a the boy would walk in $\frac{1}{2}$ h $1\frac{1}{2}$ h $2\frac{1}{2}$ h
b the train would travel in $\frac{1}{4}$ h $1\frac{1}{2}$ h $2\frac{1}{4}$ h.

Now read the distances from the graphs and see if the answers are correct.

26

Decimal fractions
tenths
hundredths
thousandths

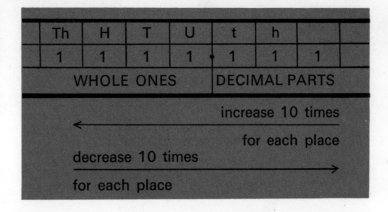

Th	H	T	U		t	h	
1	1	1	1 ·	1	1	1	

WHOLE ONES   DECIMAL PARTS

increase 10 times
for each place

decrease 10 times
for each place

Look at the diagram. Remember the decimal point separates the **whole ones** from the **parts**.

The value of the 1 (or any other figure) increases **10 times** for each place to the **left** decreases **10 times** for each place to the **right**.

The **first place** of decimals are **tenths**.
The **second place** of decimals are **hundredths**.
You must now discover the **third place** of decimals.

Whole one or unit   a   b   c▫

From the diagram, find

1 How many squares marked **a** are there in the whole one?

2 How many strips marked **b** are there in the whole one?

3 How many small squares **c** are there in strip **b**, in square **a**?

4 Without counting, how many small squares are there in the whole one?

5 What fraction of the whole one is 1 small square?

6 Write

U	t	h	th
0 ·	1		
0 ·	0	1	
0 ·	0	0	1

Square **a** is $\frac{1}{10}$ of the whole one   or

Strip **b** is $\frac{1}{10}$ of $\frac{1}{10} = \frac{1}{100}$ of the whole one   or

Square **c** is $\frac{1}{10}$ of $\frac{1}{10}$ of $\frac{1}{10} = \frac{1}{1000}$ of the whole one or

7 On graph paper draw four whole ones or units like the one shown above.
Mark them **a b c d**.
On **a** shade 500 thousandths
on **b** shade 50 thousandths
on **c** shade 5 thousandths
on **d** shade 720 thousandths.
Write each of the above as a decimal fraction.

8 Now write as a decimal fraction that part of each whole one which is unshaded.

9 How many
a tenths in 1 whole one
b hundredths in 1 tenth
c hundredths in 1 whole one
d thousandths in 1 hundredth
e thousandths in 1 tenth
f thousandths in 1 whole one?

A

Write in full each of these numbers. The first example is done for you.

	T	U	t	h	th	
1		0 ·	7	5	3	= 7 tenths, 5 hundredths, 3 thousandths
2		3 ·	0	7	9	
3	2	6 ·	5			
4	5	9 ·	0	6		
5		8 ·	3	0	4	

Draw similar columns and write these numbers.

6 Thirty-seven point three
7 Six point nought two four
8 503 tenths
9 107 hundredths
10 49 thousandths
11 703 thousandths
12 1 032 hundredths
13 2 056 thousandths
14 7 thousandths
15 5 907 thousandths

B

Write as decimals.

1 a 39 tenths   b 209 tenths   c 4 tenths   d 350 tenths

2 a 57 hundredths   b 9 hundredths   c 206 hundredths

3 a 70 hundredths   b 780 hundredths   c 1 035 hundredths

4 a 19 thousandths   b 503 thousandths   c 9 thousandths

5 a 30 thousandths   b 700 thousandths   c 2 004 thousandths

6 a 3 tenths 9 thousandths   b 6 hundredths 2 thousandths

7 a $4\frac{7}{10}$ b $6\frac{3}{100}$ c $2\frac{307}{1000}$ d $1\frac{69}{1000}$

C

1 How many tenths in   a 7·6   b 10·3   c 59?

2 How many hundredths in   a 5·03   b 0·96   c 10·5?

3 How many thousandths in   a 0·752   b 0·58   c 0·8
  d 1·725   e 3·6   f 5·54?

In each of these numbers write the value of

4 the figure 7   a 17·6   b 7 503   c 1·07

5 the figure 3   a 0·403   b 30·87   c 56·38

6 the figure 5   a 510·9   b 4·705   c 106·5

7 Write these four groups of numbers in order of size, putting the largest first.
  a 1·1   10·1   1·01   11·1   b 2·34   4·32   24·3   42·3
  c 0·83   0·803   0·8   0·82   d 7·6   7·06   7·652   7·65

8 Which number is midway between each of these pairs of numbers
  a 3·5 and 3·9   b 1·5 and 2   c 0·92 and 1·02?

9 Put a decimal point in each of these numbers so that the figure 6 becomes 6 thousandths.
  a 946   b 3 276   c 36   d 6   e 506

Decimal fractions
tenths hundredths thousandths

A Remember the rules for **multiplying** and **dividing** by 10, 100, 1 000.

H	T	U	t	h	th
	3	5	8	1	
1	3	7	2		
7	4	6			

U	t	h	th
3	5	8	1
1	3	7	2
0	7	4	6

←—10 ×—
←—100 ×—
←—1 000 ×—

Figures move **one place** to **left**.
Figures move **two places** to **left**.
Figures move **three places** to **left**.

Write the answers only.

1 a 1·723 × 100   b 90·3 × 10   c 0·681 × 1 000
2 a 0·054 × 10   b 0·32 × 1 000   c 59 × 100
3 a 0·68 × 1 000   b 176·5 × 10   c 89·37 × 10

H	T	U	t
5	8	9	7
1	2	8	3
3	0	4	

÷ 10 —→
÷ 100 —→
÷ 1 000 —→

H	T	U	t	h	th
5	8	9	7		
1	2	8	3		
0	3	0	4		

Figures move **one place** to **right**.
Figures move **two places** to **right**.
Figures move **three places** to **right**.

Write the answers only.

4 a 407 ÷ 10   b 96·4 ÷ 100   c 509 ÷ 1 000
5 a 87 ÷ 100   b 2 063 ÷ 1 000   c 0·65 ÷ 10
6 a 900 ÷ 1 000   b 3·4 ÷ 100   c 0·14 ÷ 10

B 1 Write the following numbers in columns.
Then find the totals of **a, b** and **c**.
a 13 tenths   56 units   129 hundredths   17 thousandths
b 9 thousandths   3 hundreds   134 tenths   79 hundredths
c 10 tens   209 hundredths   3 407 thousandths   105 tenths

2 Find the difference between
a 89 tenths and 137 hundredths   b fifty and 23 thousandths.

3 Complete the following then write each answer as a decimal.
a 9 tenths × 6 = ☐ tenths   b 7 hundredths × 8 = ☐ hundredths
c 1·7 × 6 = 17 tenths × 6 = ☐ tenths
d 35 tenths ÷ 7 = ☐ tenths   e 95 hundredths ÷ 5 = ☐ hundredths
f 5·4 ÷ 9 = 54 tenths ÷ 9 = ☐ tenths

4 Write the answers only.
a 2·2 × 9   b 0·8 × 7   c 13·4 × 3   d 20·7 × 5
e 0·07 × 6   f 0·14 × 8   g 1·37 × 4   h 5·35 × 6

5 Write the answers only.
a 13·8 ÷ 3   b 9·1 ÷ 7   c 2·4 ÷ 8   d 27·6 ÷ 6
e 15·48 ÷ 4   f 8·75 ÷ 5   g 2·43 ÷ 9   h 0·35 ÷ 7

Study the table below. It shows many facts you must
remember about the DECIMAL NUMBER SYSTEM.

Th	H	T	U	t	h	th
1 000	100	10	1	0·1 $\frac{1}{10}$	0·01 $\frac{1}{100}$	0·001 $\frac{1}{1000}$
10×10×10 10^3	10×10 10^2	10×1		$\frac{1}{10}$	$\frac{1}{10}×\frac{1}{10}$	$\frac{1}{10}×\frac{1}{10}×\frac{1}{10}$

Metric measures
length

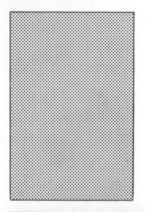

1 Measure in mm as accurately as you can
  a the length, width and diagonal of the rectangle.
     Find its perimeter.
  b the lengths of the sides of the triangle. Find its perimeter.
  c the radius of the circle. Find its diameter.

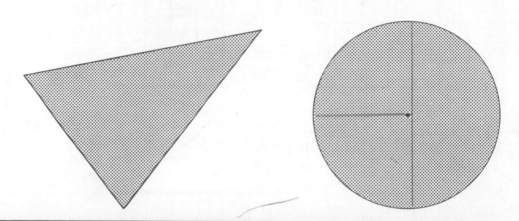

B

1 a Measure these lines in mm.

X ————————————————————————————————————

Y ——————————————————————————————————

Z —————————————————————————————————

  b Write each measurement in cm.
  c Now write each measurement as a decimal
     fraction of 1 metre.

2 Write the following lengths first in mm, then in
   metres
  a a line 10 times as long as X
  b a line 9 times as long as Y.

3 Write the following lengths first in mm, then in cm
  a a line $\frac{1}{10}$ the length of Y
  b a line $\frac{1}{3}$ the length of Z.

C 1 Find by measuring
  a the length of your pace to the nearest cm
  b the length of your shoe in mm
  c the length of your span in mm
  d your height to the nearest cm
  e the height you can reach to the nearest cm.

   Keep a record of these measurements which will
   help you to estimate distances and heights.

2 Practise making estimates of length at every
   opportunity. Compare your estimates with the
   measured distances or heights.

   Here are some things to do at home.
   You can think of many others.

   Estimate and measure

  a the length of your bed
  b the height of the table
  c the length, width and height of the living room
  d the height of your house
  e the length and width of the street in which you
     live.

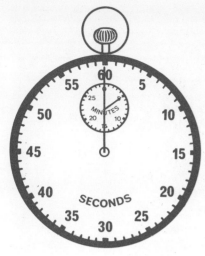

Metric measures
distance time speed

A

You will require a stop watch. Make sure that you can use it correctly. Work with a partner, taking turns, one doing the activity, the other measuring the time.

1 From a starting line walk briskly for 30 seconds. Mark the distance and measure it to the nearest metre.

2 At this rate find how many metres you can walk in
   a 1 minute   b 10 minutes.
   c Suppose you live 5 minutes' walk from school find the distance in metres.

3 Walk briskly from school to home.
   a How long does the journey take?
   b Find the distance in metres.

B

1 Measure and mark a distance of 100 metres.

2 Find the time in seconds it takes you
   a to walk 100 metres b to run 100 metres.

3 At this rate how long will it take you
   a to walk 1 kilometre   b to run 1 kilometre?
   (1 km = 1 000 metres)

4 From a book of records find the fastest times of world champions to run
   a 100 m   b 1 km.

5 Find a place which is approximately 1 km from school. With the help of your teacher choose a route which is safe and free from traffic dangers.
   a How long does it take you to walk 1 km?
   b Compare your answer with 3a.
     Give reasons for the difference in the times.

6 Find the time taken
   a to run 1 km   b to cycle 1 km.

C

1 This is a picture of a familiar road sign.
   It tells the motorist the maximum speed in km/h on that stretch of road.
   a Find other examples of the sign and write the given speed limits.
   b Find the speed limit in km/h on motorways.

2 A car travels at an average speed of 60 km/h.
   a How far will it travel in
     $\frac{1}{2}$ h $1\frac{1}{2}$ h   2 h   20 min?
   b The motorist left home at 11.30 and arrived at the end of his journey at 14.45.
     How far had he travelled?

3 A journey by air takes $1\frac{3}{4}$ h.
   The distance travelled is 700 km.
   a How far does the aeroplane travel in 15 min?
   b Find its average speed in km/h.

4 Mr Jones has to make a journey of 160 km.
   He estimates that he can travel at an average speed of 40 km/h.
   a How long will the journey take?
   b If he has a meeting at 14.30, what is the latest time at which he must leave home?

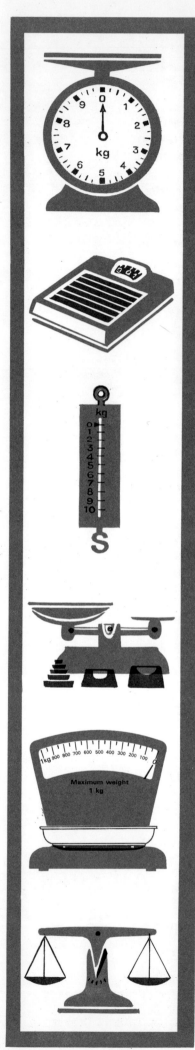

Metric measures
weight

A

There are many kinds of scales which are used for weighing. Make sure that you know how to use them to weigh accurately.

1 Collect several things of different weights e.g. parcels of books, stones, bricks, etc. Estimate the weight of each and then find by weighing the actual weight to the nearest 50 grammes.

2 Look in the food cupboard at home and make a list giving the net weight of the contents of various packets or tins, e.g. sugar, tea, tins of fruit, etc.

3 By weighing find how many medium sized potatoes weigh 1 kg.

4 Use the school weighing machine to find your own weight and that of your partner to the nearest $\frac{1}{2}$ kg.

Find
a the total of your weights
b the difference between your weights.

B

The table gives the weight in grammes of each of the given coins.

Bronze coins			Cupro-nickel coins	
$\frac{1}{2}$p	1p	2p	5p	10p
1·78 g	3·56 g	7·12 g	5·65 g	11·30 g

1 a How many $\frac{1}{2}$p coins for £1·00?
   Find the weight of £1·00 worth of $\frac{1}{2}$p coins.
  b How many 1p coins for £1·00?
   Find the weight of £1·00 worth of 1p coins.
  c How many 2p coins for £1·00?
   Find the weight of £1·00 worth of 2p coins.

2 Mark the answers to a b c.
  What do you notice about these answers?
  See if you can discover a relationship between the value and the weight of the bronze coins.

3 In the same way find the weight of £1·00 worth of
  a 5p coins   b 10p coins.
  Is there a relationship between the value and the weight of these cupro-nickel coins?

C

The measure for very heavy loads, e.g. coal, gravel, etc. is called a METRIC TONNE.
Find out all you can about the metric tonne and how loads of this weight are weighed.

Looking ahead
Using metric measures you have been finding the weight of different things both light and heavy. In future the word **mass** is likely to be used instead of **weight**. Try to find the reason for this change.

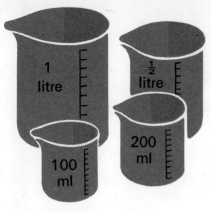

Metric measures
capacity

A

1 Use the metric measures to find, as accurately as you can, how much various containers hold in litres or millilitres, e.g. a milk bottle, a cup, a large lemonade bottle, a kettle, a bucket, a small bath.

2 Collect other containers of different sizes. Estimate and then measure how much each will hold.

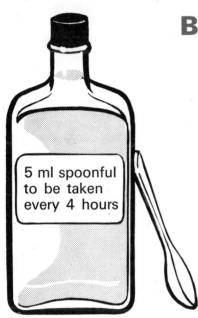

5 ml spoonful to be taken every 4 hours

B

1 Bottles for medicine are provided by the pharmacist in these six sizes.

500 ml	300 ml	200 ml	150 ml	100 ml	50 ml

What decimal fraction of a litre does each bottle contain?

2 A 5 ml spoon is given with each bottle. How many spoonfuls of medicine are contained in each bottle?

3 How long would a 150 ml bottle of medicine last if 1 spoonful was taken 3 times a day?

4 How long would a 300 ml bottle of medicine last if 2 spoonfuls were taken 3 times a day?

5 Use the 5 ml spoon to measure the capacity of various small containers, e.g. an egg cup.

C

It may surprise you to learn that the metric measures of **weight and capacity** came from a metric measure of **length**.

This is a picture of a **cube** the sides of which measure **1 centimetre**.

It is called a **cubic centimetre** which is written 1 cm³.

The weight of 1 cm³ of water is 1 gramme.

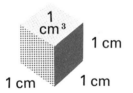

1 Write the weight 1 000 cm³ (1 litre) of water
   a in grammes   b in kilogrammes.
   Check your answer by weighing 1 litre of water.

2 How many millilitres are there in 1 litre?

3 What is the weight of 1 millilitre of water?

4 Get a measuring jar marked in cm³.
   a Measure the capacity of the various small containers used in example **B5**.
   b Measure 100 cm³ of water and fill the medicine bottle which will hold this amount exactly.

5 Write the following first as millilitres and then as litres
   a 500 cm³ b 200 cm³ c 2 000 cm³ d 1 150 cm³.

REMEMBER The weight of 1 litre of water or 1 000 cm³ of water = 1 kg (1 000 g)

Metric measures

Study this table.

It shows how the **metric system of weights and measures** is based on the **decimal number system.**

Notice that the measures which are in general use and which you have learned to use are printed boldly in colour.

Thousands 1 000	Hundreds 100	Tens 10	Units 1	tenths $\frac{1}{10}$ or 0·1	hundredths $\frac{1}{100}$ or 0·01	thousandths $\frac{1}{1000}$ or 0·001
kilometre (km)	hectometre	decametre	**metre** (m)	decimetre	**centimetre** (cm)	**millimetre** (mm)
kilogramme (kg)	hectogramme	decagramme	**gramme** (g)	decigramme	centigramme	milligramme
kilolitre	hectolitre	decalitre	**litre** (l)	decilitre	centilitre	**millilitre** (ml)

(1 litre = 1 000 cm³)

By putting in a decimal point write

1 in metres **a** 1 500 mm **b** 2 800 mm **c** 700 mm.

2 in litres **a** 1 500 ml **b** 2 600 ml **c** 300 ml.

3 in km **a** 1 500 m **b** 1 250 m **c** 800 m.

4 in kg **a** 1 500 g **b** 3 000 g **c** 650 g.

5 Write
a 1 570 m to the nearest km
b 8 010 mm to the nearest m
c 1 350 g to the nearest kg
d 6 800 ml to the nearest l.

6 Some sizes of sheets of notepaper are given in the table.

Size	Millimetres
B5	176 × 250
C5	162 × 229
B6	125 × 176
C6	114 × 164

a Write the measurements in cm.
b On newspaper draw and cut out sheets of each size.
c Get some used envelopes of different sizes.

By folding fit each of your sheets into an envelope of convenient size.

7 The measurements of a caravan as given in a catalogue are
length 4 250 mm, width 2 830 mm, height 2 500 mm.
Write each measurement in metres.

8 The following are the running distances in certain athletic events
a 1 000 m **b** 5 000 m **c** 10 000 m **d** 20 000 m.
Write these distances in km.

9 Find in km the total distance run by a relay team of 4 men each running 800 m.

10 The weights of four parcels are 850 g, 730 g, 290 g and 370 g. Find the total weight
a in grammes   **b** in kg.

11 A small barrel holds 48 litres of water.
What is the weight of the water in kg?

12 How many cm³ are there in
a $\frac{1}{2}$ litre   **b** 1·5 litres   **c** 4 litres?

13 How many metres in
a 2·7 km   **b** 1$\frac{3}{4}$ km   **c** 4$\frac{1}{2}$ km?

14 How many mm in
a 6 m   **b** 3·5 m   **c** 5·125 m?

Metric measures in the shops

Price of material per metre	
A	£3·85
B	£4·60
C	£3·25

A

1 Wood is sold in lengths of 5 metres costing 28p per length. A carpenter requires 27 m to do a job.
   a How many lengths must he order?
   b How much will the wood cost?

2 Mother wishes to make 4 curtains each needing 136 cm of material.
   a How much curtain material does she require?
   b The material can be bought in metres and half-metres only. What length must she order?
   c Find its cost at £1·80 per metre.

3 Good quality woollen cloth costs £4·20 per metre. Find the cost of
   a 50 cm   b 10 cm   c 30 cm   d 70 cm.

4 a Which is the top quality material, A, B or C?
   b Find the cost of these lengths
      $2\frac{1}{2}$ metres   B
      4 metres 20 cm   C
      3·60 metres   A.
   c The shopkeeper will cut material to the nearest 10 cm. If he receives an order for 2·95 m what length will he cut? How much will the order cost for each quality?

B

1 500 grammes of Scotch salmon cost £1·00.
   Find the cost of 1 kg.
   Find the cost of
   a 100 g   b 200 g   c 50 g   d 20 g   e 10 g.

2 Beef steak costs 50p per $\frac{1}{2}$ kg. Find the cost of 1 kg.
   Find the cost of
   a 100 g   b 200 g   c 50 g   d 20 g   e 10 g.

3 Mark and correct the answers above, then use them to find the cost of
   a 1 200 g salmon   b 300 g steak
   c 800 g salmon   d 650 g steak.

4

Price per					
1 kg	500 g	200 g	100 g	50 g	20 g
		10 p			
			3p		
	10p				
				$\frac{1}{2}$p	
5p					

   a Copy this table and make a **ready reckoner** by filling in the empty spaces.
   b Use the ready reckoner to find the cost of
      600 g cabbage @ 5p per kg
      800 g tomatoes @ 15p per $\frac{1}{2}$ kg
      1 050 g apples @ 10p per $\frac{1}{2}$ kg.

C

1 Which size of bottle is the best value for money? Give the reason for your choice.

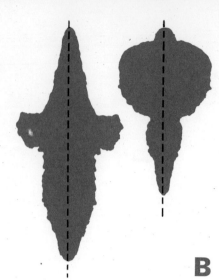

Shapes which balance

A

1 Fold a sheet of paper in half.

2 Open it and on the inside make blots of ink or paint.

3 Quickly refold the paper and press it firmly.

4 Open the paper again and look at the pattern you have made. Notice that it is the same on each side of the fold line.

5 Make several patterns in this way.
Do they all balance about the fold line?

B

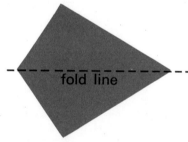

1 Fold a piece of paper in half by a vertical fold line.

2 On one side of the paper draw a shape which fits along the fold line.

3 Cut out the shape through both thicknesses of the paper.

4 Now open the paper. Look at the shape you have made. One half of the shape is the same size and looks the same as the other. The shape balances about the fold line.

C

1 Fold another piece of paper in half by a horizontal fold line.

2 On one side of the paper draw a shape which fits along the fold line.

3 Cut out the shape through both thicknesses of the paper.

4 Now open the paper.
Look at the shape you have made.
a Is one half of the shape the same size as the other?
b Does one half of the shape look the same as the other?
c Does the shape balance about the fold line?

D

In the same way by folding and cutting make shapes like these.

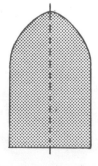
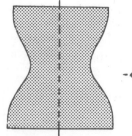
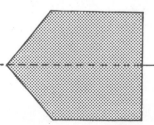
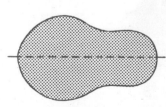

36

Shapes which balance
symmetry

The fold line about which a shape balances has a special name. It is called **a line of symmetry.**

Shapes which balance are called **symmetrical shapes.**

A

1 The shapes below are symmetrical. Draw them on squared paper and then show the line of symmetry in each case.

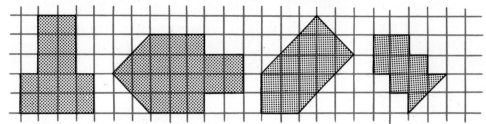

2 These are half-shapes and each line of symmetry is shown by a dotted line. On squared paper draw the complete symmetrical shape for each.

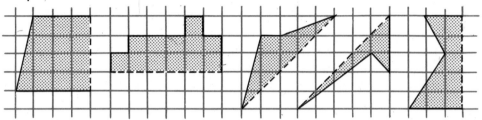

B Many shapes of leaves, plants, insects and birds are symmetrical.

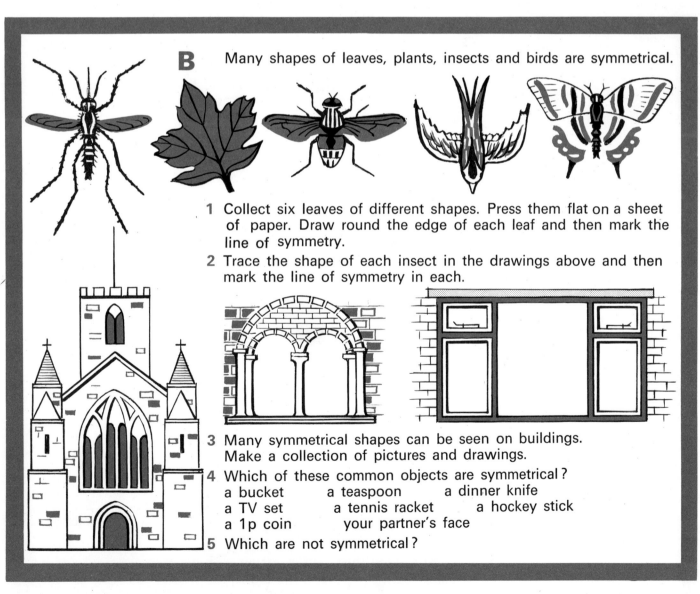

1 Collect six leaves of different shapes. Press them flat on a sheet of paper. Draw round the edge of each leaf and then mark the line of symmetry.

2 Trace the shape of each insect in the drawings above and then mark the line of symmetry in each.

3 Many symmetrical shapes can be seen on buildings. Make a collection of pictures and drawings.

4 Which of these common objects are symmetrical?
a bucket   a teaspoon   a dinner knife
a TV set   a tennis racket   a hockey stick
a 1p coin   your partner's face

5 Which are not symmetrical?

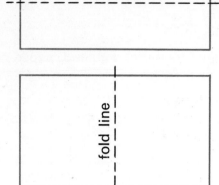

Symmetry

Some shapes have more than one line of symmetry.

A

1 Use a ruler and set square to draw two large rectangles of different sizes. Cut them out.
2 Fold each rectangle as shown in the diagrams so that the fold line is a line of symmetry. How do you know?

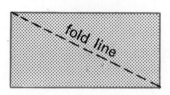 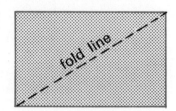

3 a Draw and cut out two more rectangles of different sizes. Fold them as shown in the diagrams.
  b In each case the fold line is a **diagonal** of the rectangle. Is the diagonal a line of symmetry? Give a reason for your answer.
4 Draw a rectangle in your book and show by dotted lines all its lines of symmetry.

B

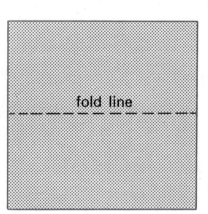

1 Draw and cut out a large square.
2 Fold it as shown in the diagram. Is the fold line a line of symmetry? How do you know?
3 The square has three more lines of symmetry. Find them by folding. Do all the lines pass through the centre?
4 Draw a square in your book and show by dotted lines all its lines of symmetry.

C

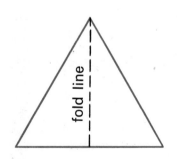

1 Measure in mm the sides of the triangle. What do you discover? Name the triangle.
2 Draw and cut out an equilateral triangle with sides twice the length of the triangle shown.
3 Fold it so that the fold line is a line of symmetry. How do you know?
4 There are two more lines of symmetry. Find them by folding.
5 Draw an equilateral triangle in your book and show by dotted lines all its lines of symmetry.

D

1 An isosceles triangle has two sides of equal length. Draw and cut out a large isosceles triangle. Find by folding how many lines of symmetry it has.
2 A scalene triangle has sides of different lengths. Draw and cut out a large scalene triangle. Find by folding the number of lines of symmetry it has.

Symmetry

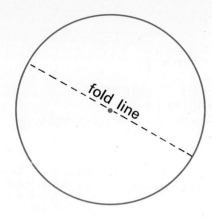

A

1 Use a pair of compasses to draw a large circle. Cut it out.
2 Try to discover by folding how many lines of symmetry there are in a circle.
3 Do all the lines of symmetry pass through the centre?
4 Each line of symmetry in a circle has a special name. What is it?
5 From what you have learnt explain why the sink or bath plug is always circular.

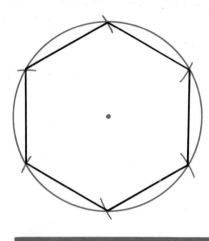

B

1 Draw a large circle and then construct a **regular hexagon.** Cut it out.
2 The hexagon has six sides. It has also six lines of symmetry. Find them by folding.
3 Draw a regular hexagon in your book and show by dotted lines all the lines of symmetry.

C Look at these shapes. Write the letters of the shapes which have

1 no line of symmetry
2 one line of symmetry
3 two lines of symmetry
4 three or more lines of symmetry.

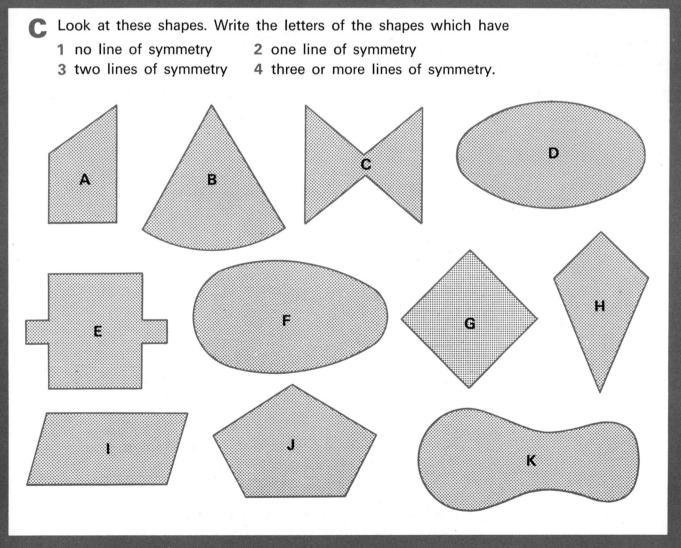

Angles

A

You have learnt that **angles** measure **turns** or **rotations**.
In a **complete** turn there are **4 right angles** or 360°.
A turn or rotation can be in either a **clockwise** or an **anti-clockwise** direction.

1 How many right angles in
   a ¼ turn   b a half turn   c ¾ turn?

2 How many degrees in
   a a right angle   b ½ right angle   c ⅓ right angle
   d 3 right angles   e a straight angle or 2 right angles?

B

1 Find the number of degrees in each angle marked *x*. Check the answers by using the set squares.

2 Which of these angles are
   a acute angles
   b obtuse angles?

3 Find the number of degrees in each angle marked *x*. Check the answers by using the set squares.
   Notice that these angles are greater than two right angles (180°) but less than four right angles (360°). Such angles are called **reflex angles**.

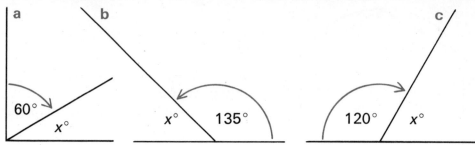

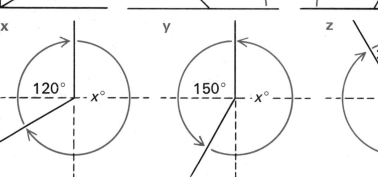

C

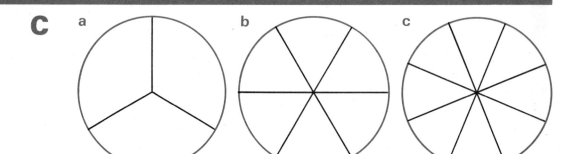

1 How many equal angles are there at the centre of each circle **a   b   c**?
  How many degrees are there in each angle?

2 a Draw three large circles. Then copy each diagram **a   b** and **c** using set squares to draw the angles at the centre.
  b Join the points on the circumference of each circle.
  c How many sides has each shape you have drawn? Name each shape.

3 a How many equal angles are there in diagram X, in diagram Y?
  b How many degrees are there in each angle in diagram X, in diagram Y?

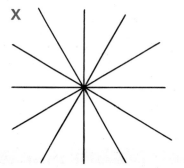

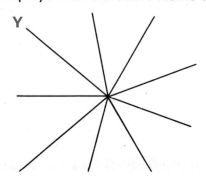

Bearings

Have you ever thought how a ship or an aeroplane is steered to its destination? There are no roads or signs at sea or in the air to guide the pilot. He must know the direction in which to steer his craft. This direction is called a **bearing** and is found by measuring **the angle in degrees from north turning clockwise.**

A

NORTH

A 90° B

1 Look at the diagram. The pilot has to steer in the direction from A to B, which is a bearing of 90°. Write this bearing in another way using a compass direction.

NORTH

B ————————— A

2 a What is the bearing measured in degrees in this diagram?
b Write the bearing in another way using a compass direction.

3 Draw the 8-point compass card. Then write the following compass directions as bearings
a South   **b** N E   **c** S E   **d** S W   **e** N W

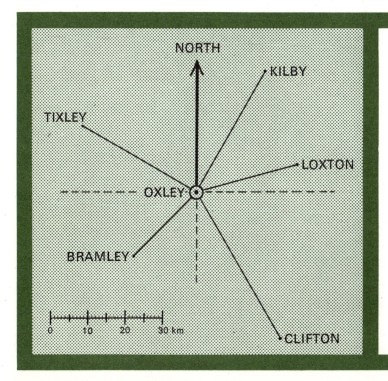

B

The map shows the distance and the direction of five towns from Oxley.
It is drawn to a scale of 1 mm to 1 000 m (1 km).

Find by measuring

1 The distance in km from Oxley to each town. (Measure from the centre of Oxley in each case.)

2 The bearing of each town from Oxley. To do this use the angles (30°, 60°, 45°, 90°) on your set squares.

3 The shortest distance in km from
a Loxton to Clifton
b Clifton to Bramley.

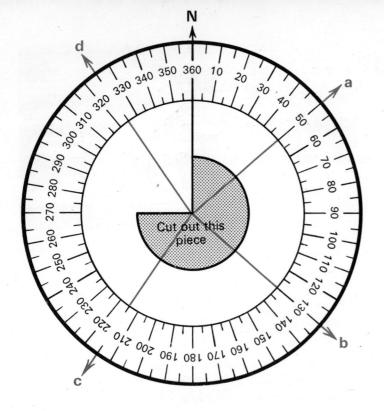

Bearings

A

Bearings are measured from the North line clockwise in any number of degrees up to 360°.

An instrument which will help you to measure and to plot bearings is called a **360° protractor.**

1 Look at this picture of a 360° protractor. Find its centre and the North line.

2 How many degrees are there in each division of the scale?

3 Read the bearings shown by the direction lines **a b c d.**

4 Find these bearings. (Place the point of your pencil on the scale.)
   a 20°   **b** 140°   **c** 225°   **d** 315°

5 You must estimate to find these bearings.
   a 33°   **b** 98°   **c** 174°   **d** 259°

B

To make a 360° protractor.

1 On thin cardboard draw a circle with a radius of 30 mm. Cut out the circle carefully.

2 Fit the circle exactly into the diagram. Hold it firmly and mark off the divisions round the edge.

3 Draw and cut out the piece at the centre, making sure that the North line and the centre of the circle are accurately placed.

4 Draw and mark a North line. Fit your 360° protractor to it. Then plot these bearings.
   a 35°   **b** 140°   **c** 175°
   d 250°   **e** 345°

5 Use your 360° protractor to find the approximate bearing from Heathrow Airport (London) to
   a Penzance
   b Dundee
   c Belfast
   d Dublin
   e Oslo
   f Copenhagen
   g Lisbon
   h Gibraltar
   i Paris
   j Amsterdam
   k Berlin
   l Milan
   m Brussels
   n Barcelona

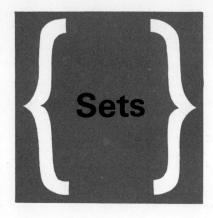

Sets

To remind you

A set is a collection of things — numbers, shapes, letters of the alphabet, people, animals, colours, games, etc.

The things which belong to a set are called **members** of the set.

The members of a set are put in **curly brackets** or **braces** and each member is separated by a comma.

Any capital letter is used to stand for a set.

Example

V = { carrot, potato, parsnip, turnip } in which case
    V stands for a set of root vegetables.

1 In this way, write the members of the following sets.
   M = the set of the months of the year beginning
       with J
   L = the set of the last six letters of the alphabet
   O = the set of the odd numbers between 8 and 28
   E = the set of the even numbers between 19 and 37
   F = the set of the factors of 24

2 Look at the members of the following sets.
   Then give a name to each set.
   A = { apple, pear, plum, raspberry, strawberry }
   C = { $\frac{1}{2}$p, 1p, 2p, 5p, 10p, 50p }
   V = { $\frac{1}{10}$, $\frac{1}{3}$, $\frac{1}{4}$, $\frac{1}{5}$, $\frac{1}{6}$ }
   N = { 15, 20, 25, 30, 35 }
   S = { △, ◣, ◬, ▽, ▷ }

The symbol

∈ stands for 'is a member of'

∉ stands for 'is not a member of'

Example
red ∈ { colours }
stands for 'red is a member of a set of colours'.
buttercup ∉ { trees }
stands for 'a buttercup is not a member of a set of trees'.

3 Write the meaning of the following
   a Susan ∈ { girls' names }
   b Terrier ∈ { dogs }
   c $\frac{19}{4}$ ∈ { improper fractions }
   d ⏢ ∉ { triangles }
   e A TEN and a FIFTY ∉ { bronze coins }

4 Write the following putting in the correct symbol ∈ or ∉ in place of ○.
   a horse ○ { animals with horns }
   b golf ○ { outdoor games }
   c $2\frac{1}{4}$ ○ { mixed numbers }
   d 20 ○ { multiples of 3 }
   e 5 ○ { factors of 30 }
   f circle ○ { triangles }

5 A set with no members is a 'null' or 'empty' set.
   a There are two ways of showing an empty set. What are they?
   b Write three examples of an empty set.

Sets and sub-sets

The Brown Family

Mr Brown Mrs Brown Peter Ann David

A

The picture shows Mr and Mrs Brown and family.

1 How many members are there in the family?

2 How many members are **a** males **b** females?

3 Write the names of the members of the set
B = { the Brown family }
You can make a diagram of this set by enclosing the names in a ring.

B
Mr Brown, Mrs Brown, Peter, Ann, David

4 Write the names of the members of the set
M = { males in the Brown family }
You can show this set by making this diagram.

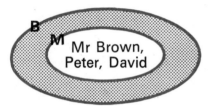

M Mr Brown, Peter, David

5 Look carefully at the members of set B and set M. Compare them. You see that all the members in set M are contained in set B. This can be shown by making this diagram.
Which names can be written in the coloured part of the diagram?

B
M Mr Brown, Peter, David

6 Draw a similar diagram to show that all the members of the set F are contained in set B.
F = { females in the Brown family }
Which names can be written in the coloured part of this diagram?

When all the members of a small set are contained in a larger set, each small set is called a **sub-set** of the larger set. So M is a **sub-set** of B, and F is a **sub-set** of B.

B

The children in Class 4 who can swim 100 metres are Susan, James, Peter, Freda, John, Mavis.

1 Draw a ring diagram to show
S = { children in Class 4 who can swim 100 metres }

S

2 Peter, Freda, Mavis can swim 200 metres.
Draw a ring diagram to show
Z = { children in Class 4 who can swim 200 metres }

Z

3 Now complete this diagram to show that Z is a **sub-set** of S.

4 Which names can be written in the coloured part of the diagram?

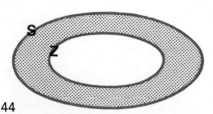

S
Z

Sets and sub-sets

A

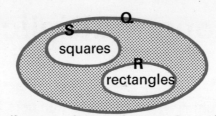

The diagram shows the set Q = { quadrilaterals }

1 Sub-sets of Q are shown in the diagram. Name them.

2 Copy the diagram and show another sub-set.

> **SYMBOLS TO REMEMBER**
> ⊂ stands for 'is a sub-set of'
> ⊄ stands for 'is not a sub-set of'

Look again at the example above. You can write

Set S ⊂ Set Q or S ⊂ Q
Set R ⊂ Set Q or R ⊂ Q

3 a Using the symbol, write the sub-sets shown in each of the following diagrams

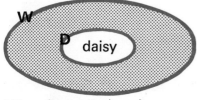

W = { wild flowers }

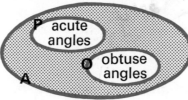

A = { angles }

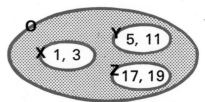

O = { odd numbers to 20 }

b Write another sub-set which could be contained in each set.

c Write the following using a symbol.
carrot is not a sub-set of W   dandelion is a sub-set of W
12 is not a sub-set of O   triangle is not a sub-set of A.

B

1 T = {10, 12, 14, 16 }
Which of the following
a is a sub-set of T   b is not a sub-set of T?
Write each answer using the correct symbol.
A = { 10 }   B = { 12, 16 }   C = { 12, 14, 16 }
D = { 8, 10, 12 }   E = { 2 }   F = { 18, 16, 14 }

2 Sets of lights are often used for signalling.
The picture shows a set of two lights. L = { red, green }
To find in how many different ways the lights can be used, the number of sub-sets must be found.

The sub-sets are R = { red }   G = { green }   Z = { red, green }
and ∅ when there are no lights.
How many ways are there in which these lights can be used?
Notice set Z is the same as set L — every set is a sub-set of itself.

The empty set is always a sub-set of a set.

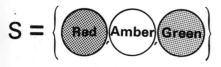

3 Here is a set of three signalling lights.
Find the number of ways in which they can be used by writing the sub-sets of S = { red, amber, green }.
Remember to include the set itself and the empty set.

4 Traffic lights consist of red, amber and green lights.
Which sub-sets of these lights are actually used?
Which other sub-sets could be used if required?

5 Write all the sub-sets of a N = { 10, 20, 30 }   b X = { L, M, O } 45

Long multiplication

$$20 = 10 \times 2$$

$$30 = 10 \times 3$$

A

1 Multiply each of these numbers by 10
   a 15   b 26   c 34   d 70   e 8·9   f 13·6.

2 To multiply a number by **20**, first multiply it by **10**, then multiply the answer by **2**.
   Multiply each of these numbers by 20
   a 14   b 28   c 0·37   d 90   e 5·8   f 16·5.

3 To multiply a number by **30**, first multiply it by **10**, then multiply the answer by **3**.
   Multiply each of these numbers by 30
   a 13   b 24   c 39   d 80   e 9·3   f 18·7.

4 Write the answers only.
   a 40 times 29   b 60 times 85   c 90 times 5·7
   d 72 × 50   e 58 × 70   f 8·3 × 30   g 4·4 × 80

5 Multiply the following and write the answers as £'s.
   a 46p × 20   b 38p × 60   c 75p × 40   d 84p × 50

6 Write the answers only.
   a £0·52 × 30   b £0·17 × 80   c £1·56 × 20   d £1·65 × 70

B

```
   37
 × 23
  740 = 37 × 20
  111 = 37 × 3
  851 = 37 × 23
```

```
    £
 1·63
 × 27
 32·60 = £1·63 × 20
 11·41 = £1·63 × 7
 44·01 = £1·63 × 27
```

Look at the given examples. Then copy the following and find the answers in the same way.

1   49
   × 35
   ___ = 49 × 30
   ___ = 49 × 5
   ___ = 49 × 35

2   68
   × 42
   ___ = 68 × 40
   ___ = 68 × 2
   ___ = 68 × 42

3   37
   × 23
   ___

4   £
   0·43
   × 26
   ___ = £0·43 × 20
   ___ = £0·43 × 6

5   £
   1·35
   × 25
   ___ = £1·35 × 20
   ___ = £1·35 × 5

6   £
   1·80
   × 32
   ___

Set down the following in the same way as questions **B3** and **B6** and find the answers.

7 27 × 34   8 61 × 53   9 89 × 27   10 56 × 76

11 £0·93 × 24   12 £1·37 × 21   13 £2·50 × 32   14 £4·25 × 28

Give the answers to the following in £'s.

15 64p × 43   16 52p × 17   17 49p × 35   18 92p × 26

C

1 Find the product of a 29 and 17   b 13 and 48.

2 Find the value of a 25^2 b 12^3.

3 Find the missing number in each of the following
   a □ ÷ 21 = 46   b □ ÷ 49 = 15.

4 David saved 25p each week. How much did he save in 1 year?

5 26 children each paid £1·34 for a school outing. How much did they pay altogether?

Long division

X	2	3	4	5	6	7	8	9
11	22	33	44	55	66	77	88	99
12	24	36	48	60	72	84	96	108
13	26	39	52	65	78	91	104	117
14	28	42	56	70	84	98	112	126
15	30	45	60	75	90	105	120	135
16	32	48	64	80	96	112	128	144
17	34	51	68	85	102	119	136	153
18	36	54	72	90	108	126	144	162
19	38	57	76	95	114	133	152	171
20	40	60	80	100	120	140	160	180
21	42	63	84	105	126	147	168	189
22	44	66	88	110	132	154	176	198
23	46	69	92	115	138	161	184	207
24	48	72	96	120	144	168	192	216
25	50	75	100	125	150	175	200	225
26	52	78	104	130	156	182	208	234
27	54	81	108	135	162	189	216	243
28	56	84	112	140	168	196	224	252
29	58	87	116	145	174	203	232	261

A

Study the table. It gives the products when numbers from 11 to 29 (in the left-hand column) are multiplied by 2, 3, 4, . . . 9 (on the top row).

Example 1
19×7
Move your finger along the line opposite 19 until it is under 7 in the top row.
$19 \times 7 = 133$

Example 2
$216 \div 27$
Move your finger along the line opposite 27 until it reaches 216. Look above in the top row for the answer which is 8.
$216 \div 27 = 8$

Find from the table

	a	b	c
1	17×8	13×9	28×5
2	26×9	24×6	23×7
3	$117 \div 13$	$90 \div 15$	$128 \div 16$
4	$162 \div 18$	$184 \div 23$	$130 \div 26$

5 Find the product of
   a 22×9 b 25×8 c 29×7

B Division with remainders

Example $140 \div 21$ or $21\overline{)140}$
Using the table above move your finger along the line opposite 21 until it reaches the number **nearest to but less than** 140. This number is 126 and above it in the top row is 6.
Subtract 126 from 140 which gives the remainder 14.

$$
\begin{array}{r}
6 \text{ r. } 14 \\
21\overline{)140} \\
126 = 21 \times 6 \\
\hline
14
\end{array}
$$

Copy each of the following and find the answer

	a	b	c	d
1	$17\overline{)147}$	$15\overline{)101}$	$18\overline{)173}$	$19\overline{)143}$
2	$25\overline{)217}$	$26\overline{)161}$	$24\overline{)200}$	$28\overline{)260}$
3	$16\overline{)139}$	$14\overline{)105}$	$27\overline{)209}$	$29\overline{)255}$

C

The examples in **B** have **units** only in the answers. Study the example
$1\,325 \div 29$
which has **tens** and **units** in the answer and is therefore worked in the two stages marked **a** and **b**.

Example   Use the table
a to divide 132 **tens**
b to divide 165 **units**

$$
\begin{array}{r}
45 \text{ r. } 20 \\
29\overline{)1325} \\
29 \times 4 \text{ tens} = 116\downarrow \\
165 \\
29 \times 5 \text{ units} = 145 \\
\hline
\text{remainder} \quad 20
\end{array}
$$

Work the following using the table.
1 $21\overline{)609}$
2 $15\overline{)945}$
3 $23\overline{)1\,357}$
4 $26\overline{)1\,352}$
5 $17\overline{)629}$

D

Some of the following have remainders.

	a	b	c	d
1	$20\overline{)615}$	$24\overline{)1\,608}$	$25\overline{)925}$	$37\overline{)853}$
2	$19\overline{)600}$	$13\overline{)377}$	$16\overline{)471}$	$22\overline{)2\,047}$
3	$15\overline{)1\,360}$	$12\overline{)900}$	$28\overline{)2\,104}$	$26\overline{)1\,583}$

47

Long division

When a multiplication table is not used you must estimate the number required to find the necessary products. It will help if a trial figure is found by dividing the number of tens in the divisor into the number of tens in the dividend. Study these examples.

$$\begin{array}{r} 4 \text{ r. } 5 \\ 23\overline{)97} \\ 92 \\ \hline 5 \end{array}$$ How many 2's in 9?   Try 4. $23 \times 4 = 92$

$$\begin{array}{r} 8 \text{ r. } 7 \\ 32\overline{)263} \\ 256 \\ \hline 7 \end{array}$$ How many 3's in 26?   Try 8. $32 \times 8 = 256$

A

Work the following.

1 $31\overline{)281}$ 2 $22\overline{)157}$ 3 $43\overline{)225}$ 4 $51\overline{)420}$

5 $34\overline{)177}$ 6 $33\overline{)209}$ 7 $41\overline{)330}$ 8 $20\overline{)197}$

In the following examples round off the divisor to the nearest 10 before finding the 'trial' figure.

9 $29\overline{)218}$ 10 $19\overline{)183}$ 11 $28\overline{)152}$ 12 $37\overline{)321}$

13 $48\overline{)402}$ 14 $18\overline{)140}$ 15 $27\overline{)155}$ 16 $38\overline{)240}$

Can you find a reason for rounding off the divisors to the nearest 10?

B

Sometimes the 'trial' figure gives a product which is too big or too small. Think what you must then do to get the correct answer. Work the following.

1 $17\overline{)130}$ 2 $26\overline{)182}$ 3 $46\overline{)419}$ 4 $24\overline{)218}$

5 $35\overline{)321}$ 6 $28\overline{)200}$ 7 $16\overline{)122}$ 8 $15\overline{)137}$

C

Refer to page 47, Section **C**. Work the following in the same way without using the table. The examples will help you.

REMEMBER
Every remainder must be less than the divisor.
9 is the largest 'trial' figure

	a	b
1	$25\overline{)585}$	$32\overline{)864}$
2	$37\overline{)666}$	$26\overline{)1\,482}$
3	$32\overline{)1\,724}$	$21\overline{)2\,075}$
4	$33\overline{)2\,673}$	$41\overline{)3\,116}$
5	$29\overline{)1\,079}$	$24\overline{)2\,187}$
6	$42\overline{)2\,772}$	$38\overline{)1\,298}$
7	$17\overline{)1\,003}$	$14\overline{)896}$

Example 1
$$\begin{array}{r} 27 \\ 29\overline{)783} \\ 58\downarrow \\ \hline 203 \\ 203 \\ \hline \end{array}$$
$58 = 29 \times 2$
$203 = 29 \times 7$

Example 2
$$\begin{array}{r} 91 \text{ r. } 11 \\ 15\overline{)1376} \\ 135\downarrow \\ \hline 26 \\ 15 \\ \hline 11 \end{array}$$
$135 = 15 \times 9$
$15 = 15 \times 1$

8 Now try the following.
Think carefully where to put the decimal point in the answer.

a $13\overline{)£5.98}$ b $15\overline{)£5.25}$ c $23\overline{)£16.56}$ d $27\overline{)£31.05}$

Multiplication and division

Car performance			
Registration number of car	Petrol litres	Distance km	km per litre
NRY51D	54	540	
BBC391F	40	520	
247KL		693	11
RBY106F	75		13
AJF408E	86	1 032	
ZXK669H		1 134	14

A 1 Find the missing measurement for each car listed in the table.

2 A car travels 11 km per litre of petrol. From Aberdeen to Edinburgh is approximately 185 km. How much petrol to the nearest litre does the motorist use for a journey from Aberdeen to Edinburgh and back?

3 The petrol tank of a car holds 40 litres. A motorist fills it at the start of a journey from Glasgow to London (630 km) and puts in a further 35 litres on the journey. If he has 30 litres left in the tank when he reaches London find
  a how much petrol he used
  b the number of km per litre.

B Find the missing numbers in each of these examples.

1
```
    □□
  × 3 9
  ━━━━
  □□□□
   □□□
  ━━━━
  1 7 9 4
```

2
```
    2 8
  × □□
  ━━━━
  □□□□
   □□□
  ━━━━
  1 5 6 8
```

3
```
        □□
  2 9)1 2 4 7
```

4
```
          8 4
  1 7)□□□□
```

5
```
          2 4
  □□)1 3 6 8
```

C 1 Paul deposits 12p each week in the Savings Bank. How much has he saved at the end of 42 weeks?

2 A shopkeeper bought a number of toys for which he paid £11·66. Each toy cost 22p. How many toys did he buy?

3 Share a £25·00 prize equally among 16 winners.
  a How much does each receive?
  b How much remains?

4 a Find how much is paid for the bicycle if 36 payments are made as advertised.
  b By how much is it cheaper to pay cash for the bicycle?

5 A transistor radio costs £13·40. Peter pays a deposit of £2·00 and the rest of the price by 12 monthly instalments.
  How much does he pay each month?

6 A length of cloth measuring 18 metres cost £25·20. Find the cost of a 1 m b 5 m c 13 m.

BICYCLES

Cash Price

£27·50 or

36 Payments of 85p

49

Area
squares and rectangles

The amount of surface contained in a shape is called its **area**.
Area is measured by counting equal squares.
Square centimetres (cm²) when the measurements are in **centimetres**.
Square metres (m²) when the measurements are in **metres**.

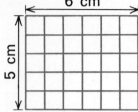

A

1 a Draw a rectangle 6 cm long and 5 cm wide.
     Mark it out in cm² as shown.
   b How many cm² in each row?
   c How many rows are there?
   d Find the area of the rectangle. Check your answer by counting cm².

2 If the rectangle had measured 6 m long and 5 m wide, what would
   have been its area?

3 Find the area of each of these rectangles. Look carefully at the
   measurements of the length and the width before writing the answer.

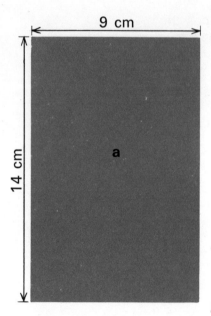

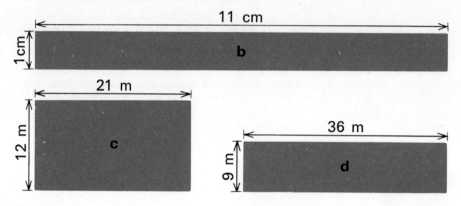

B

1 Find the area of each of the following squares or rectangles.
   a 20 cm by 4 cm   b 19 cm by 10 cm   c 15 m by 8 m
   d 24 cm square   e 35 m by 26 m   f 86 m by 3 m

2 Find the area of each of these rectangles. Remember ½ = 0·5.
   a 9½ m by 5 m   b 13·5 m by 8 m   c 7 m by 10·5 m
   d 15 m by 7·5 m   e 23·5 cm by 10 cm f 84 cm by 2·5 cm

3 Find the area of a plot of land 35 m long and 25 m wide.

4 Which bedroom in a house has the greater area and by how many m²
   Bedroom 1   4 m by 3·5 m   or   Bedroom 2   5 m by 2·5 m?

5 a How many tiles 10 cm square will cover 1 square metre?
   b How many tiles would be needed to cover a surface 5 m by 3 m?

C

1 Measure to the nearest cm the length and width of

   a the front cover of this book   b a sheet of school writing paper
   c a single sheet of newspaper   d the top of your desk.
   Then find the area of each in cm².

2 Measure to the nearest metre the length and breadth of the classroom.
   Find its area in m².

From the examples above you learn that
AREA = Number of units (length) × Number of units (breadth)
or as the formula A = l × b = lb.

Area
squares and rectangles

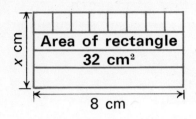

A

Look at the diagram. You are told the **area** of the rectangle and also its **length**.

1 Which measurement (*x*) is there to find?
To find this measurement answer these questions.

2 a How many cm² are there in the area of the rectangle?
b How many centimetre squares are there in one row?
c How many rows of centimetre squares are required to give 32 altogether?
d What is the breadth of the rectangle?

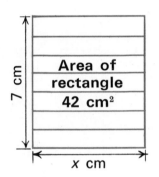

B

1 Look at this diagram.
a Which measurements are given?
b Which measurement is there to find?

2 What is the area of the rectangle?

3 How many rows of centimetre squares are there?

4 How many centimetre squares are there in each row?

5 What is the breadth of the rectangle?

C

1 Find the answers to the following.
Check your answers by making drawings.
a A rectangle has an area of 54 cm².
Its breadth is 6 cm. Find its length.
b A rectangle has an area of 72 cm².
Its length is 12 cm. Find its breadth.
c A square has an area of 100 cm².
Find its length and breadth.

2 A rectangle measures 6 cm by 4 cm.
a Draw a rectangle of the same area which is 8 cm long. How wide is it?
b Draw a rectangle of the same area which is 2 cm wide. How long is it?
c Can you draw a square of this area?
Give a reason for your answer.

3 a Write the measurements of all the rectangles you can draw having an area of 36 cm².
b Can you draw a square having an area of 36 cm²?

4 Which of the following are the areas of squares?
a 72 cm² b 64 cm² c 81 cm² d 50 cm² e 49 cm²

5 Write a sentence to say
a how the **length** of a rectangle can be found when the **area** and **breadth** are given.
b how the **breadth** of a rectangle can be found when the **area** and **length** are given.

6 Write the sentences **5a** and **5b** as formulae using **A** for area, **l** for length and **b** for breadth.

Area
squares and rectangles

A

Find the missing measurement in each of the following.

1

	area	length	breadth
a	27 cm²		3 cm
b	45 cm²	9 cm	
c		13 cm	8 cm
d	27·5 m²		5 m

2

	area	length	breadth
a	64 cm²		8 cm
b	180 m²	18 m	
c	81 m²		9 m
d		25 cm	7 cm

3

	area	length	breadth
a		6·5 m	9 m
b		13·5 m	7 m
c		10 m	8·5 m
d	67·5 cm²		5 cm

B

Shape W has been made into two rectangles by the dotted line.

1 Find **a** the area of each rectangle
   b the area of the shape.

2 Draw a sketch of shape W and make it into two rectangles in another way.

3 Find its area. Compare this answer with **1b**.

4 a Make a sketch of shapes X and Y.
   b Show by a dotted line how each shape can be made into two rectangles.
   c Find the area of each shape.

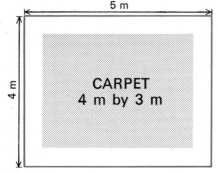

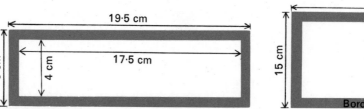

C

The carpet is laid in the middle of the floor of a room leaving a surround uncovered.

1 Find the area of
   a the room   **b** the carpet.

2 What is the area of the surround?

3 How wide is the surround?

4 Find the area of the border in each of the following.

5 A photograph measures 28 cm by 12 cm.
It is mounted on a piece of cardboard to leave a border 2·5 cm wide all round.

Find **a** the length and width of the cardboard
   b the area of the photograph
   c the area of the cardboard
   d the area of the border.

6 A lawn measures 27 m long and 26 m wide.
A rose bed 16 m by 9 m is cut out.
Find the area of **a** the rose bed   **b** the grass.

Area and perimeter
squares and rectangles

A

In some workshops **length** measurements are made in **millimetres** so areas are found in **square millimetres (mm²)**.

1 Look at the drawing.
It shows 1 square centimetre divided into square millimetres.
   a How many mm in 1 cm?
   b How many mm² in 1 cm²?

1 cm²

2 a Measure in mm as accurately as possible the length and breadth of each of the shapes X, Y, and Z.

X

Y

Z

   b Find the area of each shape in mm².
   Check your answers.
   c Which shape has the largest area?
   d Which has the smallest area?
   e By how many mm² is the largest area greater than the smallest?

3 a Each square which makes up this pattern has an area of 25 mm².
   What does each side of one square measure?
   b Find the length and breadth of the rectangle in mm.
   c What is the area of the rectangle in mm²?
   d If all the squares were cut out and arranged to make one large square, what would be the measurement of its sides?

B

The perimeter of a shape is the distance all round it. The distance is measured in **units of length (millimetres, centimetres or metres)**.

1 Find the perimeter of
   a a square of 16 cm sides
   b a rectangle 9·5 m by 7·5 m
   c a rectangle 97 mm by 163 mm.

2 Find the perimeter of each of the shapes X, Y, Z in Section **A2** above.

3 Find the perimeter and area of rectangles measuring
   a 23 cm by 14 cm   b 8·5 m by 12 m
   c 65 mm by 40 mm.

4 The perimeter of a square is 40 cm. Find
   a the length of each side   b the area of the square.

5 The perimeter of a rectangle is 138 mm.
   Its length is twice its width.
   Find a the length and the width
        b the area of the rectangle.

6 Write a sentence to say how the perimeter of
   a a square   b a rectangle
   can be found.

Measuring angles

A

1 Get a 60° set square and a 45° set square.
   Write in degrees
   a the angles you can draw or measure using the set squares
   b the angles up to 180° you can draw or measure by combining the angles on the set squares.

2 Look at these angles carefully.
   Choose the angle you think measures
   45°, 75°, 105°, 120°, 135°, 165°.

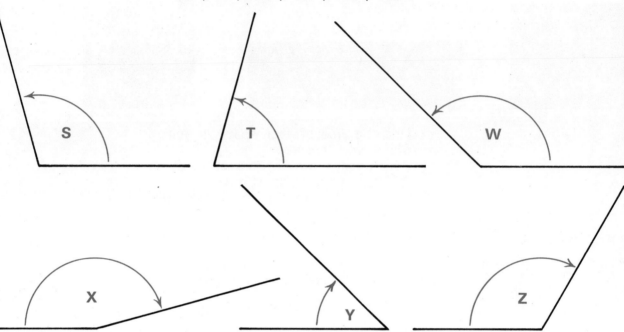

Check your answers by measuring each angle using the set squares.

3 Explain why it is impossible by using the angles on the set squares to draw or measure angles of
   a 65°   b 140°.

B

From the exercises above you will note that angles of certain sizes only can be measured with the set squares. To measure angles of any size in degrees you will most likely use a **180° protractor**. You can make and learn to use the simple protractor shown in the diagram.

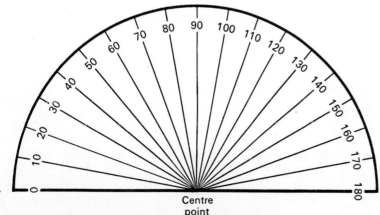

1 On stiff paper draw a semi-circle of 45 mm radius and cut it out. Mark the centre.

2 Fit the semi-circle exactly into the diagram keeping the base lines level and one centre point over the other. Mark off the divisions round the edge.

3 Join each of the division points to the centre point as shown.

4 Number each division line carefully. How many degrees are there in each division?

5 Up to how many degrees will this protractor measure?

Measuring angles

A

Using the protractor

Measuring angles in a **clockwise** direction.

1 Acute angles

Place the protractor on the angle so that the point marked P is exactly on V (the vertex of the angle).

The line 0° to 180° must be exactly over the arm of the angle. Read from the scale the number of degrees. What is the size of the angle?

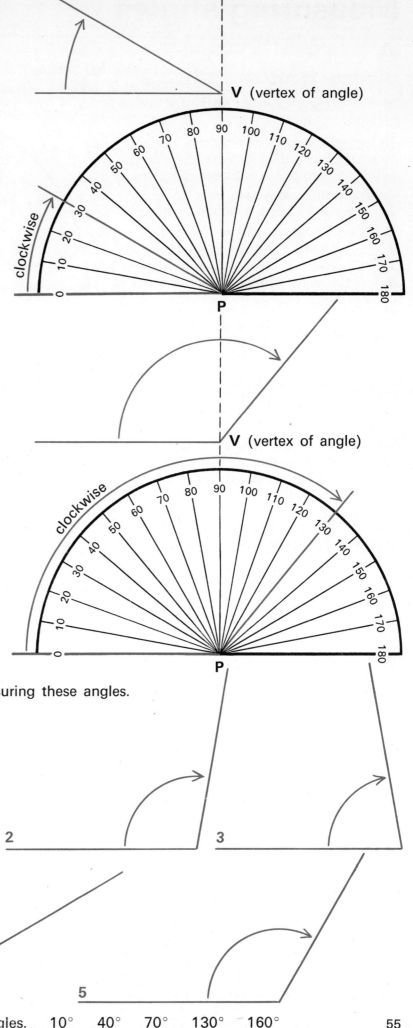

V (vertex of angle)

2 Obtuse angles

Do exactly the same again to measure the given obtuse angle.
What is the size of this angle?

V (vertex of angle)

B Practise using your protractor by measuring these angles.

1

2

3

4

5

6 Measure clockwise and draw these angles.   10°   40°   70°   130°   160°

55

Measuring angles

A

Using the protractor

Measuring angles in an **anti-clockwise** direction.

To measure angles in an anti-clockwise direction, place the protractor on the angle as before. Look at the diagram.

1 Counting anti-clockwise how many 10° divisions does the angle measure?
2 Write this measurement in degrees.

   You can see that to measure angles anti-clockwise it would be easier if the protractor was numbered anti-clockwise.

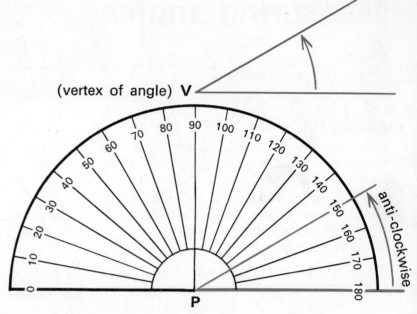

3 Draw a chart in your book like the one below and write in the angles which correspond. Notice that each pair of angles adds up to 180°. Give a reason for this.

0°	10°	20°	30°	40°	50°	60°	70°	80°	90°	100°	110°	120°	130°	140°	150°	160°	170°	180°
180°	170°	160°	150°	140°	130°	120°	110°	100°	90°	80°	70°	60°	50°	40°	30°	20°	10°	0°

4 Now mark on your protractor the corresponding angles so that there are two numbers for each division as shown on the diagram.

5 What is the size of each angle
   a measured anti-clockwise
   b measured clockwise?

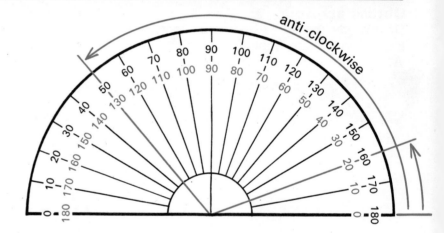

B Practise using your protractor by measuring these angles.

1

2

3

4

5

6 Measure anti-clockwise and draw these angles. 30° 50° 80° 100° 120° 150°

The protractor measuring and drawing angles

A

The instrument shown in the drawing is a protractor marked to measure angles of every degree from 0° to 180°. It is made of clear plastic so that the lines of the angle can be seen through it. Obtain such a protractor from your teacher and examine it. Notice that the degrees are numbered in tens

0° 10° 20° up to 180°

a clockwise on the outer edge
b anti-clockwise on the inner edge.

You must learn to count the degrees in between the tens.

1 Is the angle AOC acute or obtuse? What is its size?
2 Is the angle BOC acute or obtuse? What is its size?

Think before measuring or drawing an angle.   Is it smaller or greater than 90°?

B Practise using the protractor by measuring these angles as accurately as possible.

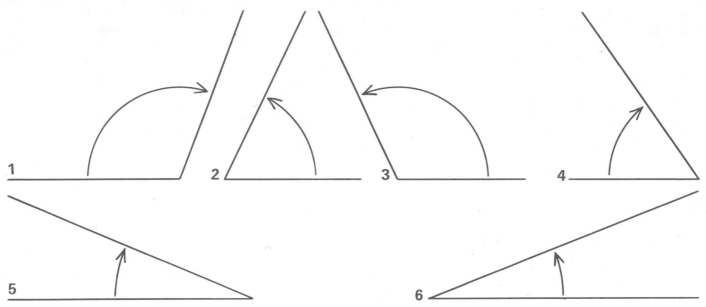

7 Draw any three acute angles and any three obtuse angles. Measure each one as accurately as you can to the nearest degree. Ask your partner to check your measurements.

8 A ——————————— B
   a Draw a number of horizontal lines of any length. Mark each one AB.
   b With A as the vertex draw angles of 5°, 23°, 48°, 75°, 127° and 164°.
   c With B as the vertex draw angles of 18°, 35°, 52°, 85°, 105° and 155°.
   Ask your partner to check your measurements.

9 Draw a number of
   a vertical lines
   b oblique lines of any length.
   Mark each one AB.
   With A as the vertex draw angles of 38°, 82°, 115°, 146°.

Non-rigid and rigid shapes

A Non-rigid shapes

Obtain some plastic or cardboard strips and fasteners. The strips need not be of the same length.

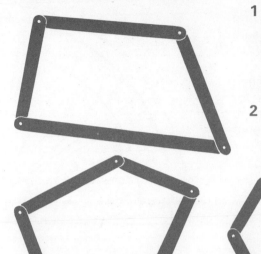

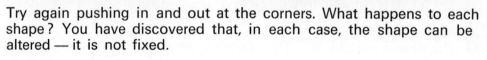

1 Take four of the strips and fasten them together to make a quadrilateral.
Push the shape in and out at the corners.
What happens to the shape?

2 Now use more strips to make these shapes
a pentagon (5 sides)
a hexagon (6 sides).

Try again pushing in and out at the corners. What happens to each shape? You have discovered that, in each case, the shape can be altered — it is not fixed.

B Rigid shapes

1 Now make several triangles, using strips of different lengths.
Try pushing each one in and out at the corners.
Write what you discovered when you tried to alter the shape.
The triangle is known as a rigid (fixed) shape because its shape can only be altered by changing the length of its sides. This fact is important to builders and engineers, e.g. in constructing roofs, bridges, pylons, etc. Collect pictures showing the use of triangles in buildings, etc.

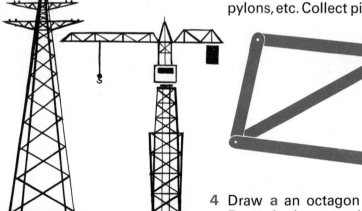

2 Take the quadrilateral again and fix a cross piece as shown in the diagram. Now try to alter its shape by pushing at the corners. What do you discover?

3 Experiment in the same way with
a the pentagon b the hexagon.
Find how many cross pieces are required to make these shapes rigid.

4 Draw a an octagon (8 sides)   b a decagon (10 sides).
Draw the least number of cross pieces to make each of these shapes rigid. How many triangles are there in each shape?

5 Draw a shape having 7 sides of different lengths (a heptagon). Divide it into triangles. How many
a cross pieces   b triangles are required?

6 Draw this table in your book and make a record of your results.

7 A simple rule connects the numbers in each line. Use the rule to find the number of
a cross pieces   b triangles
in a 12-sided shape (a duodecagon).

For many centuries this method of dividing any straight line shape into triangles has been used when measuring the area of fields and plots of land.

Shape		Number of cross pieces	Number of triangles
Name	No. of sides		
triangle	3	0	1
quadrilateral	4		
pentagon	5		
hexagon	6		
heptagon	7		
octagon	8		
decagon	10		

Triangles
sides and angles

A

The chart will help you to remember important facts which you have discovered previously about different kinds of triangles.

Triangles are named
a from their sides
b from their angles.

REMEMBER
The sum of the angles of any triangle is 180°.

Equilateral triangle	Isosceles triangle	Scalene triangle
60° 60° 60°		
3 equal sides 3 equal angles (60°)	2 equal sides equal angles opposite equal sides	Sides and angles of different sizes

Acute-angled triangle   each angle is acute (less than 90°)
Obtuse-angled triangle   one angle is obtuse (more than 90°)
Right-angled triangle   one angle is 90°

1 **a** Measure in mm the lengths of the sides in each of the triangles below.
   b Name each triangle from its sides.

2 **a** Find by calculation the number of degrees in each angle marked *x* or *y*.
   b Check your answers by measuring the angles with a protractor.
   c Name each triangle from its angles.

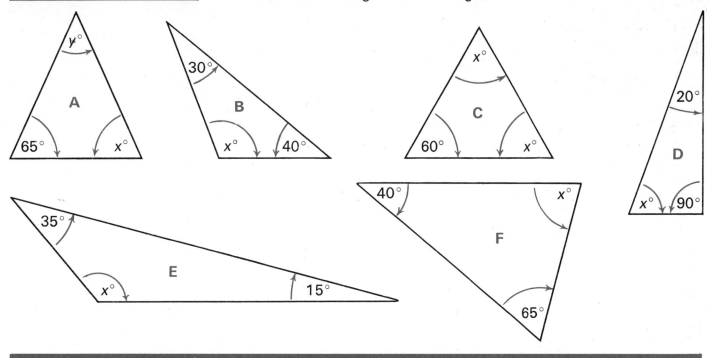

B

The drawings will remind you how to construct triangles when the lengths of the sides are given. Compasses are used to draw the intersecting arcs.

1 Measure in mm the base of each triangle.
2 Measure the other two sides in each triangle.
3 Draw triangles with these sides

a	110 mm	90 mm	67 mm
b	62 mm	20 mm	70 mm
c	89 mm	36 mm	89 mm.

4 Name each triangle according to
   a its sides   **b** its angles.

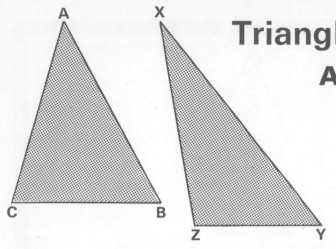

Triangles sides and angles

A Triangles and other shapes are marked by a capital letter at each vertex.
The symbol △ is short for 'triangle'.
The symbol ∠ is short for 'angle'.

1 In the △ ABC
  a Measure in mm the sides AB, BC and AC.
  b Use a protractor to measure ∠ACB, ∠ABC and ∠CAB.
  c Which angle is opposite the longest side, the shortest side?

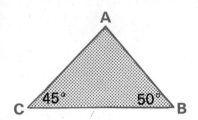

Sides		Angles	
XY		XZY	
YZ		ZXY	
ZX		XYZ	

2 a In the same way measure the sides and angles of △XYZ. Draw this table in your book and write in the measurements.
  b Which side is opposite the largest angle?
  c Which side is opposite the smallest angle?
In each example you find that the largest angle is opposite the longest side and the smallest angle is opposite the shortest side.

3 Check if this is true by looking at all the triangles on page 59.

B

1 a Draw a base line 80 mm long.
  b At each end of the base line draw ∠ACB = 45° and ∠ABC = 50°.
  c Complete the triangle and measure its sides.
  d Find ∠CAB.

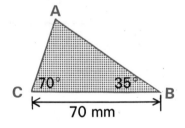 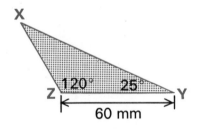

2 a Draw each of the triangles to the given measurements.
  b Measure in mm the lengths of the sides.
  c Find the size of the third angle in each triangle.

C

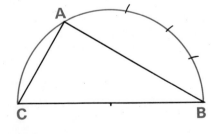

1 a Draw a semi-circle at least 120 mm in diameter.
  b Mark a point A on the circumference and join it to B and C.
  c Use a set square to measure **the angle at the circumference** (∠BAC).

2 Mark other points on the circumference and join them to B and C. Then measure these angles at the circumference. What do you discover about them?

3 Draw other semi-circles. What do you find out about the angles at the circumference?

Right– angled triangle

an ancient discovery

A Centuries before the birth of Christ, the Egyptians found a method of making a right-angled triangle. This was a most important discovery because the right angle was essential in their building work and in the measuring of land (surveying).

You can read about the Egyptian 'rope stretchers' in a library book.

The following exercise shows you how they went to work — try it in the playground. The Egyptians, of course, used different units of measurement from ours.

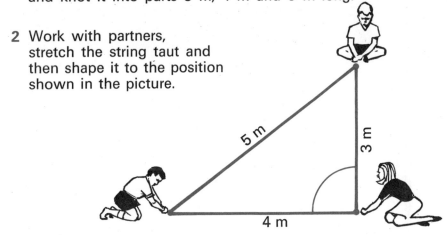

3 m   4 m   5 m

1 Get a length of thick string or rope 12 m long and knot it into parts 3 m, 4 m and 5 m long.

2 Work with partners, stretch the string taut and then shape it to the position shown in the picture.

3 Keep the string stretched and steady whilst another partner marks the right angle on the floor with chalk.

4 Do this again several times in different places taking turns to mark and test the right angles with a large set square.

5 Use this method to check all the right angles on a netball court or football pitch.

6 Find out which tools are used by builders or carpenters to make sure that walls are at right angles with the floor or windows and doors are set square.

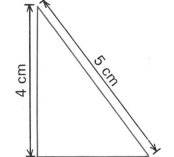

B 1 a Using ruler and compasses draw a triangle with sides 3 cm, 4 cm and 5 cm.
   b Check the right angle with a set square.

2 This special triangle is sometimes called the 3, 4, 5 triangle.
Draw triangles with sides
   a twice this length   b half this length.

3 In each case what kind of triangle have you drawn?

4 In the 3, 4, 5 triangle
the shorter sides are 3 and 4 the longest is 5.
   a Write the value of $3^2 + 4^2$, the value of 5^2.
   b What do you discover?

5 Do this again using the right-angled triangle of sides 6, 8, 10 cm. What do you discover?

6 Use this method to find which of these are right-angled triangles
   a 2, 3, 7 cm   b 4, 8, 9 cm   c 5, 12, 13 cm.

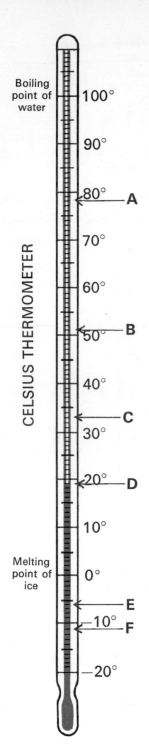

Boiling point of water — 100°

90°

80° — A

70°

60°

CELSIUS THERMOMETER

50° — B

40°

30° — C

20° — D

10°

Melting point of ice — 0°

— E

−10° — F

−20°

Measuring temperature
Celsius thermometer

A

The thermometer is used to measure temperatures in degrees Celsius (°C).

It is a sealed glass tube which is partly filled with a liquid, usually alcohol, or mercury. The liquid rises in the tube as the temperature increases and falls as the temperature decreases. The temperatures are measured on the scale marked alongside the tube.

1 Look at the drawing showing the Celsius scale. Write the temperature
  a at which water boils
  b at which ice melts or water freezes.

2 How many degrees are there between the boiling point and the freezing point of water?

3 How many degrees does each small division on the thermometer scale represent?

4 A minus sign (−) shows temperatures below freezing point. How many degrees below freezing point can be measured on this thermometer?

5 Write the temperatures at the points marked A, B, C, D, E, F on the scale. Check your answers.

6 By how many degrees does the thermometer rise or fall from
  a A to B b D to C c E to D d E to F?

B

Get a Celsius thermometer like the one shown in the drawing.

1 Find the temperature of
  a water as it comes from the cold tap
  b warm water in which you can wash yourself. Ask your teacher if you may do this
  c the water used to make tea.

2 a Measure the temperature of the air in your classroom at these times.
    09.30   12.00   15.00
  b Do the temperatures remain reasonably constant? Why?
  c If the temperatures vary widely, find a reason.
  d What is the approximate temperature of a room in which you can sit comfortably?

3 Find the difference between the inside and outside temperatures at each of the times given in **2a**.

4 Find the temperatures in the sun and in the shade in the playground. What is the difference between them?

5 To cook a chicken Mother has to heat the oven to 170°C. How many degrees above the boiling point of water is this?

6 A doctor uses a special short thermometer to measure body temperature. How does he do this? Normal body temperature is 36·9°C. A boy has a fever and his temperature is 39°C. By how many degrees is his temperature above normal?

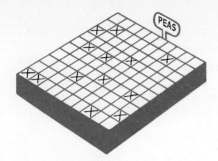

Percentages

A

This is a picture of a seed pan in which pea seeds have been sown, one in each section.

1 How many seeds have been sown?

2 The crosses show which seeds did not germinate (grow). Complete
   a ☐ seeds out of 100 seeds germinated
   b ☐ seeds out of 100 seeds did not germinate.

3 What fraction of the seeds
   a germinated   b did not germinate?

4 Write these answers as decimal fractions.

5 Out of 100 broad bean seeds which were sown 91 germinated. What fraction of the total
   a germinated   b did not germinate?

6 Write these answers as decimal fractions.

B

	Possible	Correct
Test B	100	70
Test C	100	78
Test D	100	63

1 Out of number test A of 100 items Susan had 55 correct. What fraction of the total did she have
   a correct   b wrong?

2 Write these answers as decimal fractions.

3 a In the same way write the results from tests B   C   D shown in the table.
   b In which test A   B   C   D did Susan get the best result?

C

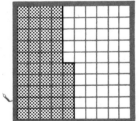

% per cent means 'out of 100'

A short way of writing 'out of 100' is **per cent** or by the sign **%**
e.g. 13 out of 100 = 13 per cent or 13%.

Write these as percentages
1 5 out of 100   2 17 out of 100   3 30 out of 100
4 1 out of 100   5 50 out of 100   6 9 out of 100
7 24 out of 100   8 84 out of 100   9 100 out of 100.

D

1 Count the number of small squares in the whole.

2 How many small squares are
   a shaded   b unshaded?

3 What fraction of the whole is
   a shaded   b unshaded?
   Now write these fractions in their lowest terms.

4 Write these answers as decimal fractions.

5 What percentage of the small squares are
   a shaded   b unshaded?

6 Write each of these fractions as a percentage, then as a decimal fraction.
   a $\frac{9}{100}$ b $\frac{27}{100}$ c $\frac{45}{100}$ d $\frac{72}{100}$ e $\frac{81}{100}$

7 Write these decimal fractions as percentages.
   a 0·53   b 0·06   c 0·68   d 0·94   e 0·12

8 Write each of these percentages as (i) a decimal fraction, (ii) a fraction in its lowest terms.
   a 7%   b 2%   c 35%   d 60%   e 1%
   f 26%   g 90%   h 100%   i 48%   j 73%

Percentages

A

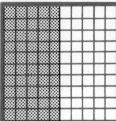

1   a Count the number of small squares in the whole of **A**.
    b How many small squares are shaded?
2   What fraction of the whole is shaded?
3   Write this fraction in its lowest terms.
4   Write $\frac{1}{2}$ (50 out of 100) as
    a a decimal fraction   b a percentage.

B

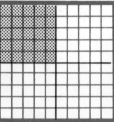

1   In **B** how many small squares are shaded?
2   What fraction of the whole is shaded?
3   Write this fraction in its lowest terms.
4   Write $\frac{1}{4}$ (25 out of 100) as
    a a decimal fraction   b a percentage.
5   Write $\frac{3}{4}$ (75 out of 100) as
    a a decimal fraction   b a percentage.

C

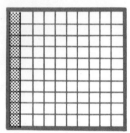

1   In **C** how many small squares are shaded?
2   What fraction of the whole is shaded?
3   Write this fraction in its lowest terms.
4   Write $\frac{1}{10}$ (10 out of 100) as
    a a decimal fraction   b a percentage.
5   Write $\frac{9}{10}$ (90 out of 100) as
    a a decimal fraction   b a percentage.
6   In the same way write each of the following as
    (i) a decimal fraction   (ii) a percentage.
    a $\frac{3}{10}$ $\frac{7}{10}$ b $\frac{1}{5}$ $\frac{4}{5}$ c $\frac{2}{5}$ $\frac{3}{5}$

A table to learn and remember

$\frac{1}{2}=0{\cdot}5=50\%$	$\frac{1}{10}=0{\cdot}1=10\%$	$\frac{2}{10}$ or $\frac{1}{5}=0{\cdot}2=20\%$
$\frac{1}{4}=0{\cdot}25=25\%$	$\frac{3}{10}=0{\cdot}3=30\%$	$\frac{4}{10}$ or $\frac{2}{5}=0{\cdot}4=40\%$
$\frac{3}{4}=0{\cdot}75=75\%$	$\frac{7}{10}=0{\cdot}7=70\%$	$\frac{6}{10}$ or $\frac{3}{5}=0{\cdot}6=60\%$
	$\frac{9}{10}=0{\cdot}9=90\%$	$\frac{8}{10}$ or $\frac{4}{5}=0{\cdot}8=80\%$

You see from this table that vulgar fractions, decimal fractions or percentages express the same fraction in a different way.

D   Find the value of
1   50% of 72   2   25% of 80   3   75% of 280
4   20% of 25   5   10% of 120   6   100% of 99
7   50% of £4·00   8   25% of 92p   9   75% of 48p
10   20% of £2·35   11   10% of 65p   12   100% of £7·00.

E   Work the following row by row.
1   10% of 190.   Now find a 30%   b 70%   c 90% of 190.
2   10% of 45p.   Now find a 30%   b 70%   c 90% of 45p.
3   20% of 245.   Now find a 40%   b 60%   c 80% of 245.
4   20% of £1·20.   Now find a 40%   b 60%   c 80% of £1·20.

Percentages

A

Find the value of the following.

	a	b	c
1	50% of 1 000	10% of 390	20% of 320
2	60% of 500	40% of 450	30% of 30
3	100% of 27½	25% of 90	70% of 250
4	20% of 75p	10% of 35p	75% of 72p
5	50% of £2·68	25% of £3·60	80% of £5·00
6	70% of £3·00	20% of £8·50	90% of £4·50

B

1 In a school 40% of the children were boys.
   a What percentage of the children were girls?
   b There were 350 children in the school.
     How many were boys? How many were girls?

2 Tony received £2·50 for birthday presents.
  He saved 20%.
   a How much did he save?
   b How much did he spend?

3 Of 850 people who attended a concert 40% were women, 10% were men and the remainder were children.
   a How many women were there? How many men?
   b What percentage of the audience was children?
   c How many children were there?

4 Decrease 90 by   a 10%   b 30%   c 70%.

5 Increase £1·60 by a 20%   b 25%   c 50%.

6 A toy priced at £2·40 was reduced by 20%.
   a By how much was the price reduced?
   b How much was paid for the toy?

7 This sale notice is the kind of advertisement which is often seen in a shop or newspaper.
   a By what % are the goods reduced in the sale?
   b Find the reduction in price of each article.
   c Find the sale price of each.

SALE All prices reduced by **10%**

	Usual Prices
Socks	45p per pair
Ties	60p each
Shirts	£1·50 and £1·80
Trousers	£7·20
Jackets	£9·90

£1·00 = 100p

C

1 Find the value of
   a 1% of £1·00   b 2% of £1·00   c 3% of £1·00.

2 Find the value of the following percentages of £1·00.
   a 5%   b 12%   c 15%   d 27%   e 3½%   f 7½%

3 What % of £1·00 is a 2p   b 9p   c 18p   d 5½p?
  When money is deposited in a bank, **interest** is paid on every £1·00 saved. Interest is usually reckoned as a percentage. Find the interest

4 at 5% on   a £1·00   b £7·00   c £23·00

5 at 2½% on   a £1·00   b £9·00   c £17·00

6 at 6½% on   a £1·00   b £12·00   c £25·00.

7 Find out the rates of interest on savings paid by
   a the Post Office Savings Bank
   b the Trustee Savings Bank c Building Societies.

8 Find out all you can about National Savings Certificates and the Contractual Savings Scheme.

65

Area triangles

A

You have learnt previously that **area** is the amount of surface contained in a shape. It is measured by counting equal squares. Up to now, the areas of squares and rectangles have been measured using square millimetres (mm²), square centimetres (cm²) or square metres (m²).

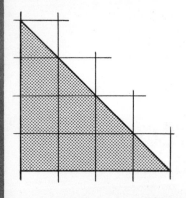
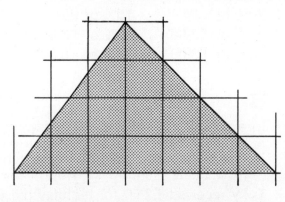

To find the areas of triangles by counting squares would be much more difficult.
Can you find a reason for this by looking at the diagrams?

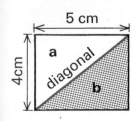

B

1 On thin cardboard draw a rectangle to the given measurements. Cut it out.

2 Draw a diagonal and cut along it.

3 You have cut the rectangle into two right-angled triangles. Fit one on top of the other. Are the triangles equal in area?

4 What is the area of the rectangle in cm²?

5 What is the area of the triangle **a**, the triangle **b**?

6 Do this exercise again using a rectangle 10 cm long and 5 cm wide. What is the area of the rectangle, the triangle **a**, the triangle **b**?

7 In the same way find the area of each of the right-angled triangles shown below.

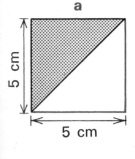

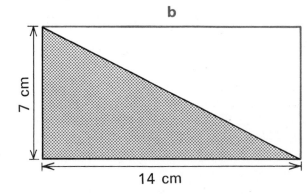

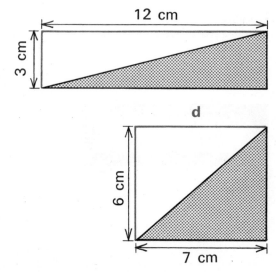

C

Find the area of each of these right-angled triangles. The two shorter sides are given. Look carefully at the measurements.

1 3 cm and 6 cm	**2** 9 cm and 6 cm	**3** 10 cm and 8 cm
4 12 cm and 10 cm	**5** 16 cm and 14 cm	**6** 20 cm and 20 cm
7 18 cm and 24 cm	**8** 12 cm and 15 cm	**9** 48 mm and 36 mm
10 55 mm and 30 mm	**11** 35 mm and 25 mm	**12** 17 cm and 11 cm

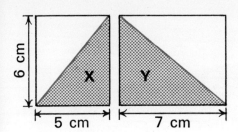

Area triangles

A

1 Draw two rectangles to the given measurements. Cut them out.

2 In each rectangle draw a diagonal and colour a right-angled triangle as shown.

3 What is the area of
 a the right-angled triangle X
 b the right-angled triangle Y?

4 Fit the two rectangles together as shown and stick them in your book.

5 You have a large triangle inside a large rectangle.
 a What is the area of the large rectangle?
 b What is the area of the large triangle?

6 Show that the area of the large triangle is half the area of the large rectangle.

7 Do this exercise again using two rectangles, the one 6 cm by 8 cm, the other 10 cm by 8 cm. Show that the area of the large triangle is half that of the large rectangle.

8 The rectangles below are the same size. Find
 a the area of each rectangle
 b the area of each triangle.
 c The triangles are of different shapes. What do you discover about their areas?

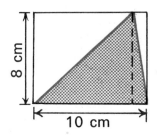 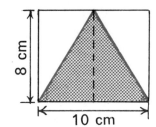 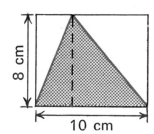

Note
(i) the base of each triangle is the same as the length of each rectangle (10 cm).
(ii) the height of each triangle is the same as the width of each rectangle (8 cm).

9 Look at the diagram.
 a What does the base of the triangle measure?
 b What does the height of the triangle measure?
 c Find the area of the triangle in cm².

REMEMBER
Area of a triangle $= \frac{1}{2}$ (base × height) $= \dfrac{\text{base} \times \text{height}}{2}$

B

Find the area of each of these triangles

1	base 12 cm	height 5 cm	2 base 21 cm	height 10 cm
3	base 11 cm	height 7 cm	4 base 10 cm	height 10 cm
5	base 15 cm	height 8 cm	6 base 19 cm	height 13 cm

Area irregular shapes

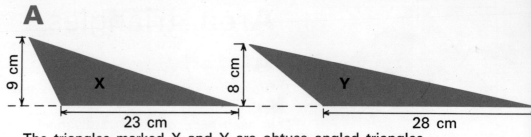

A

The triangles marked X and Y are obtuse-angled triangles.

1 Write the length of the base of **a** triangle X **b** triangle Y.
2 Write the height of **a** triangle X **b** triangle Y.
3 Find the area of each triangle in cm².
4 Which triangle has the greater area and by how many cm²?

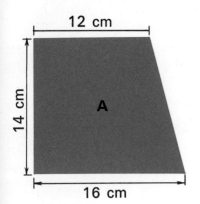

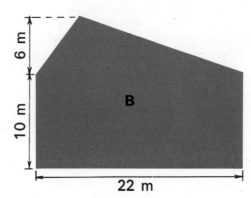

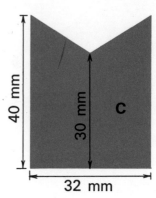

B
1 Make sketches of shapes A, B and C. Show by a dotted line how each shape can be made into a triangle and a rectangle.
2 Find the area of each shape. (Look carefully at the measurements.)

C

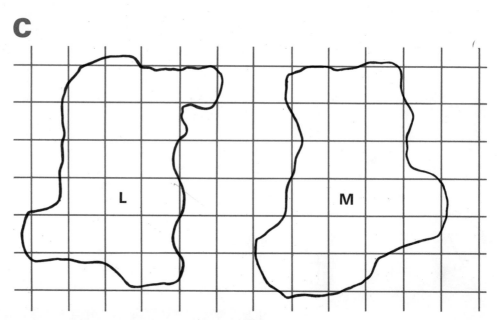

1 Look at these irregular shapes. Estimate which has the greater area.
2 Now count the whole squares in each shape. Count as whole squares those which are **a half or more.** Do **not** count the squares which are **less than a half.**
3 Each square is 1 cm². Measure some of them to see if this is correct.
4 Find the approximate area in cm² of each shape L and M.
5 Place your shoe on a sheet of paper ruled in cm². Draw round the edge. Then by counting squares find its approximate area.
6 In the same way find the approximate areas of
   a your hand with fingers closed
   b large leaves from different kinds of plants or trees, e.g. cabbage, horse-chestnut, etc.

Fixing position co-ordinates

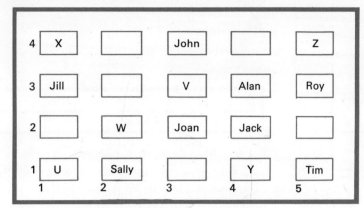

This is the plan of a classroom showing the children's desks arranged in columns and rows. Starting from the bottom left-hand desk, the columns are numbered across from 1 to 5, the rows are numbered up from 1 to 4. John's desk is (3, 4) that is 3 across and 4 up.

1 Name the children who sit in these positions.
   a (4, 2)   b (5, 3)   c (1, 3)   d (3, 2)
   e (5, 1)   f (4, 3)   g (2, 1)
   The pairs of numbers which give the position are called **co-ordinates**. The **co-ordinates** are put in brackets with a comma between them.

2 Write the co-ordinates of the desks marked U, V, W, X, Y and Z.

B

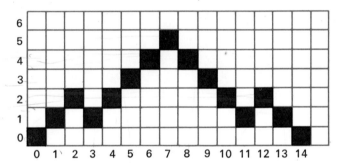

This pattern has been made by shading squares on squared paper. Starting with 0 in the bottom left-hand corner, the squares have been numbered **across** from 0 to 14 **up** from 0 to 6.

1 Write in order the co-ordinates of each shaded square.

2 Make up a pattern by shading squares. List the co-ordinates and give the list to your partner. See if he can make your pattern without seeing it by using the co-ordinates.

C

On this grid the lines have been numbered because you are going to fix the position of points.

1 Write the co-ordinates of these points.
   A, B, C, D, E, F, G, H

2 What are the co-ordinates of the point 0 (nought)?

3 On squared paper mark the lines on a grid like the example. Plot these points.
   (2, 7)   (4, 3)   (5, 5)   (6, 3)   (8, 7)
   Join the points in order by straight lines.
   What letter have you drawn?

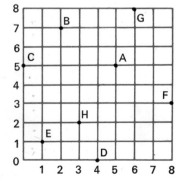

69

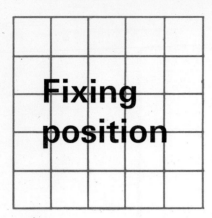

co-ordinates

A

1 Draw this table in your book. Fill in the missing numbers.

0	1	2	3	4	5	6	7	8	9	10
0	6	12	18	24						

2 Look at the pairs of numbers (0, 0) (1, 6) (2, 12) etc. and find the relationship between each pair.

3 Get a large sheet of graph paper, draw the horizontal and vertical axes and mark the scales as shown in the diagram.

4 Plot the points (0, 0) (1, 6) (2, 12) (3, 18) etc. Then draw the graph making the line as long as the paper will allow.

5 Give a reason why the graph is a straight line.

B

Extend the scales on both axes then find these answers from the graph.

1 9 sixes   2 12×6 3 14×6

4 16×6 5 How many sixes in 42?

6 $24 \div 6$ 7 $78 \div 6$ 8 $90 \div 6$

9 $3\frac{1}{2} \times 6$ 10 $8\frac{1}{2} \times 6$ 11 $12\frac{1}{2} \times 6$

12 $15 \div 6$ 13 $39 \div 6$ 14 $57 \div 6$

Check your answers by calculation.

C

1 In the same way draw the graph of the table of 8's.

2 Set yourself some examples in multiplication and division by 8 and find the answers from the graph. Check the answers by calculation.

D

These shapes are squares.

1 Draw this table in your book and complete it by finding the perimeters of squares of the given sides.

Length of side (units)	1	2	3	4	5	6
Perimeter of square (units)						

2 Look at the pairs of numbers. Find the relationship between the length of the side and the perimeter of a square.

3 On a large sheet of graph paper draw the horizontal and vertical axes and mark the scales as before extending them as required.

4 Plot the points and draw the graph. Why is it a straight line graph?

5 From your graph find the perimeters of squares of a 8 cm   b 10 cm   c 13 cm.

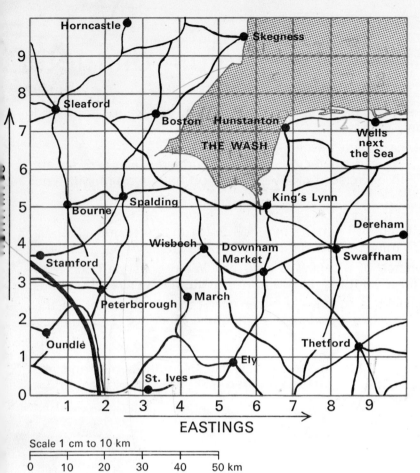

Scale 1 cm to 10 km

0   10   20   30   40   50 km

EASTINGS

Fixing position
map

A

This is a road map showing part of the east coast of England. From a map in your atlas find

1 its position on the coast

2 the names of the English counties shown

3 the name of the sea.

The map is covered with a grid, the lines of which are numbered
0 to 9 for eastings
0 to 9 for northings.

B

1 To find the position of towns on the map the co-ordinates are given
e.g. Bourne is (1, 5).
Find Bourne on the map.

2 Many towns, however, are not situated on the lines but within a square. In this case the co-ordinates of the bottom left-hand corner of the square are given, e.g. March is (4, 2).
Find March on the map.

3 Find the names of the towns in these positions.
a (0, 1)   b (2, 5)   c (2, 9)   d (6, 5)   e (9, 7)
f (0, 7)   g (3, 0)   h (5, 0)   i (4, 3)   j (0, 3)

4 Write the co-ordinates of these towns
a Peterborough b Boston c Skegness d Swaffham e Thetford.

C

Find the approximate distance 'as the crow flies'
1 from Peterborough to   a Boston   b Thetford

2 from Ely to   a Sleaford   b Stamford.

3 a Measure several of the squares. Are they the same size?
b What is
the length and width of one square, the area of one square?
c Find the actual measurements represented by
the length and the width in km of one square,
the area of one square in km².

4 By counting across squares find the approximate distances in km from
a Boston to St. Ives   b Stamford to Swaffham
c Bourne to King's Lynn.

5 By counting squares find the approximate area in km² of the sea shown on the map.

71

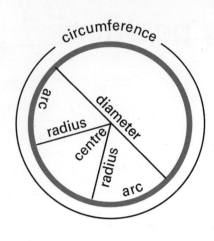

Circles
circumference radius diameter

A

The diagram shows the parts of a circle which you know from work done previously.

1 Write a sentence which describes
 a a radius   b a diameter   c the circumference   d an arc.
This is a difficult exercise. Talk over what you have written with your teacher.

2 Use compasses to draw circles of
 a radius   30 mm 45 mm 23 mm. Write the diameter of each circle.
 b diameter 70 mm 50 mm 84 mm. Write the radius of each circle.

3 The circumference of a wheel is 31·4 cm. What distance does the wheel travel for a 10 turns b 100 turns? Write the answers in metres.

4 The groundsman has to mark a circle of 10 metres diameter on the playing field. He has a wooden peg and a ball of string. Describe how he does it.

B

Practise using your compasses to draw circles, semi-circles and arcs accurately. Copy these patterns and then make up some of your own.

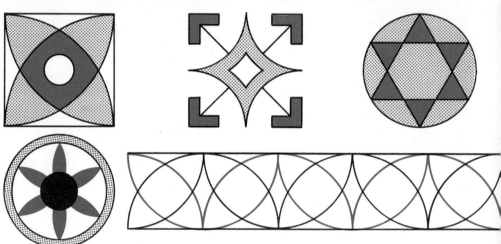

C

1 Draw a hexagon inside a circle of 50 mm radius.

2 Divide the hexagon into 6 triangles. Mark each angle at the centre x.

3 Cut out the triangles and fit them one over the other.

4 What do you discover about
 a the sides of the hexagon   b the angles at the centre of the circle?

5 What is the size of each angle at the centre?
Use a protractor to check the answer.

6 Count the number of sides and the number of triangles in each of these shapes. Calculate the size of the angles at the centre.
Enter the results in the table.

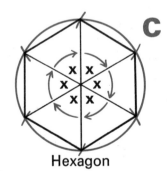

Hexagon

Pentagon

Octagon

Shape	No. of sides	No. of triangles	Angle at centre
Hexagon	6	6	$\frac{360}{6} = \square°$
Pentagon			
Octagon			

7 Fill in the table for a a decagon (10 sides) b a duodecagon (12 sides).
A shape with more than **four sides** is called a **polygon**.
If all the sides and angles are equal it is called a **regular polygon**.

To measure the **diameter** of a circle

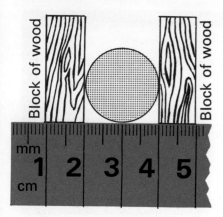

Block of wood | Block of wood

mm
1 2 3 4 5
cm

To measure the **circumference** of a circle

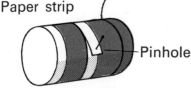

Paper strip
Length of string
Pinhole

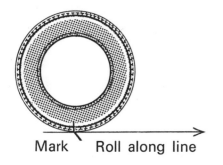

Mark   Roll along line

Circles
circumference diameter

A

Study the drawings which will remind you how to measure
the **diameter** of a circle
the **circumference** of a circle.

1 On a piece of thin cardboard draw and cut out a circle with a diameter of 10 cm.
2 Measure to the nearest cm its circumference.
3 How many times approximately is the circumference greater than the diameter?
4 Draw and cut out a circle with a diameter of 7 cm. Measure to the nearest cm its circumference.
5 Divide the length of the circumference by the length of the diameter.
   Write and complete
   $\frac{\text{circumference}}{\text{diameter}}$ or $\frac{c}{d} = \frac{\Box \text{ cm}}{\Box \text{ cm}}$
   $\frac{c}{d}$ is $\Box$ times approximately.

> **REMEMBER**
> **The answer tells you the circumference is that NUMBER OF TIMES greater than the diameter.**

B

Circular object	c	d	$\frac{c}{d}$
Cardboard circle		10cm	
circle		7cm	
New penny			
2p coin			

1 Obtain a new penny. Measure in mm
   a its circumference   b its diameter.
2 Find how many times the circumference is greater than the diameter. Write
   $\frac{c}{d} = \frac{\Box \text{ mm}}{\Box \text{ mm}}$ is $\Box$ times approximately.
3 In the same way measure a 2p coin and find the approximate value of $\frac{c}{d}$.
4 Collect several circular objects of different sizes, e.g. tin lids, jars, plates, wheels, etc.
   a Measure the circumference and the diameter of each to the nearest cm or in mm according to size.
   b Find for each object the approximate value of $\frac{c}{d}$.
5 Draw a table like the one above and enter all the results. If you have worked carefully you will find that the circumference of all the circles is a little more than 3 times the diameter.
   This is an important **relationship** or **ratio** which is usually reckoned as 3·14 (to the nearest second place of decimals) or less frequently as $3\frac{1}{7}$.
   It is called by the Greek letter π (pi).

73

Circles

circumference radius diameter

A

1 Write a sentence telling how to find the circumference of a circle when
a the diameter is given   b the radius is given.

Using $\pi = 3\cdot14$, find the length of the circumference of a circle if the diameter is

2 a 1 metre b 10 metres c 7 metres d 12 metres.
Check the answers and write them to the nearest $\frac{1}{2}$ metre.

3 a 6 cm   b 10 cm   c 40 cm   d 16 cm.
Check the answers and write them to the nearest cm.

4 a 10 mm b 30 mm   c 50 mm   d 80 mm.

5 Find the length of the circumference of a circle if the radius is
a 4 cm   b 35 mm   c 1 metre d 9 metres.
Write the answers to the nearest whole unit.

6 Using $\pi = 3\frac{1}{7}$ find the length of the circumferences of circles of the following measurements
a diameter 7 cm   b radius 7 m
c diameter 49 mm   d radius $3\frac{1}{2}$ m.

B

1 The wheels on the bicycle measure 30 cm in diameter. How far does the bicycle travel when the wheels turn round once? Use $\pi = 3\cdot14$.

2 If the wheels turned 100 times how far has the bicycle travelled?
Write the answer to the nearest metre.

3 If the wheels turned 1 000 times, how many metres less than 1 km has the bicycle travelled?

4 Measure in cm the diameter of a motor car wheel. Find the distance the car travelled in
a 1 turn b 100 turns c 1 000 turns of the wheels.

C

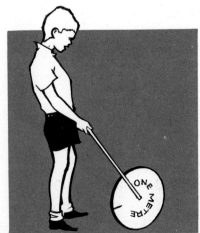

A trundle wheel is useful to measure the circumferences of large circles.

1 Take a 1 metre trundle wheel on to the playground or playing field and use it to measure circles and semi-circles marked on the games pitches.

2 Check your answers by measuring each diameter and multiplying it by π.

Plans drawing to scale

A

1 Measure this line in mm.

_____

What length does the line represent using the following scales?
a 1 mm to 10 mm (1 : 10)   b 1 mm to 5 mm (1 : 5)
c 1 mm to 20 mm (1 : 20)

2 Measure this line in mm.

_____

What length does the line represent using the following scales?
Write the answers first in mm, then in m.
a 1 mm to 100 mm (1 : 100)   b 1 mm to 50 mm (1 : 50)
c 1 mm to 200 mm (1 : 200).

A table to remember
1 000 mm = 1 m
500 mm = 0·5 m
100 mm = 0·1 m
50 mm = 0·05 m
10 mm = 0·01 m

3 The scale **1 mm to 1 000 mm** (1 : 1 000) can be written **1 mm : 1 m**.
In this way re-write the following scales.
a 1 mm to 500 mm (1 : 500)   b 1 mm to 100 mm (1 : 100)
c 1 mm to 200 mm (1 : 200)   d 1 mm to 10 mm (1 : 10)
e 1 mm to 20 mm (1 : 20)   f 1 mm to 50 mm (1 : 50)

Using the given scales draw lines to represent the following lengths.
4 1 mm to 10 mm (1 : 10)   a 300 mm   b 750 mm   c 1·5 m
5 1 mm to 50 mm (1 : 50)   a 800 mm   b 1 m   c 2·5 m
6 1 mm to 0·1 m (1 : 100)   a 1·5 m   b 7·5 m   c 12·6 m
7 1 mm to 0·5 m (1 : 500)   a 9 m   b 25 m   c 52·5 m

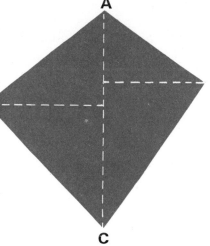

A

B

C

B

1 This is a plan of a field drawn to the scale 1 mm to 1 m (1 : 1 000).
  a Using this scale what is the actual length in m represented by 10 mm, by 7 mm, by 38 mm?
  b Measure the sides of the plan and find the actual measurements of AB, BC, CD, DA.
  c Find the perimeter of the field.
  d Copy the plan to the same scale. Measure the dotted lines and use a set square to draw them.

School corridor
Scale 1 mm to 0·1 m (1 mm to 100 mm) 1 : 100

2 The plan of the school corridor is drawn to the given scale.
  a Find its actual length and width in metres.
  b What is the area of the corridor in m²?
  c Draw the plan to the scale 1 mm to 0·05 m (1 mm to 50 mm) 1 : 50.

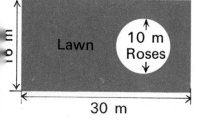

Lawn   10 m Roses

30 m

3 From the measurements given
  a draw a plan of the lawn and the rose bed to the scale 1 mm to 0·2 m (1 : 200)
  b find the perimeter and the area of the whole of the garden.

4 Using the given scales draw plans of the following.
  a Scale 1 mm to 10 mm (1 : 10)   a shelf 1 000 mm by 300 mm
  b Scale 1 mm to 0·05 m (1 : 50)   a window 2·5 m by 1·5 m
  c Scale 1 mm to 0·1 m (1 : 100)   a square lawn of 15 m sides.

Plans drawing to scale

A

Maps are drawn to many different scales according to the extent of the area represented by the map.

This scale is taken from a road map.

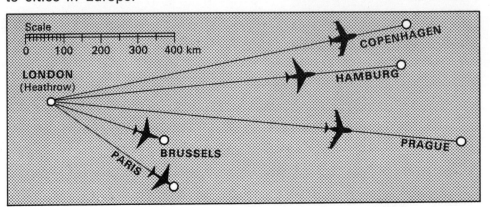

```
|||||||||||
0    5    10    15    20 km
```

1 Find the distance represented on this scale by
   a 10 mm (1 cm)   b 1 mm   c 5 mm.

2 What distances are represented by
   a 30 mm   b 50 mm   c 90 mm
   d 45 mm   e 75 mm   f 118 mm?

3 Draw scales like that above showing
   a 10 mm (1 cm) to 10 km
   b 10 mm to 20 km.

4 Find and make a note of different scales used in an atlas and other books of maps.

B

The map shows some BEA flights from London (Heathrow Airport) to cities in Europe.

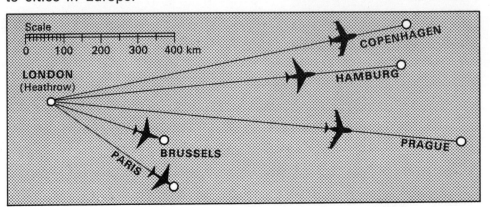

1 Find from your atlas the countries in which these cities are situated.

2 Look at the scale then find the distance in km represented by
   a 10 mm   b 1 mm   c 15 mm   d 25 mm.

3 By measuring and using the scale find the approximate flying distance from Heathrow Airport to each of the cities.

C

Plans and elevations

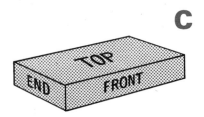

1 a Obtain a small box and measure in mm its length, breadth and height.
   b Draw full size a plan of the box. Use a set square to draw the right angles.
   c Hold up the box so that the **front** is facing you and is level with your eyes. Draw full size what you see. Mark the drawing **front elevation.**
   d Hold up the box again so that the **end** is facing you and is level with your eyes. Draw full size what you see. Mark the drawing **end elevation.**

> **REMEMBER**
> A **plan** shows shape and size seen from **above.**
> An **elevation** shows shape and size seen from **the front or end.**

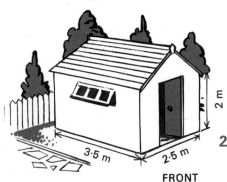

2 The picture shows the dimensions of a garden shed. Its full height is 3 m. Use the scale 1 mm to 100 mm (1 : 100) and draw
   a the plan of the shed
   b the front elevation (omit the door)
   c the end elevation (omit the window).

Venn diagrams

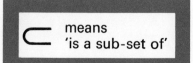
A

1 Draw a large diagram like the one shown.

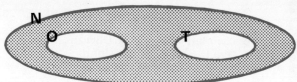

N = { whole numbers between 18 and 32 }
O = { odd numbers between 18 and 32 }
T = { whole numbers between 18 and 32 divisible by 10 }.

2 Put the correct numbers in the set and sub-sets in your diagram.

3 Use the symbol ⊂ to show the sub-sets of N.

B

1 F = { 0·7, 0·49, 0·01, 0·11, 0·98, 0·55, 0·05, 0·08, 3·0 }
Draw a large diagram and write the decimal fractions in the correct sub-sets.

M = { fractions more than $\frac{1}{2}$ but less than 1 }
T = { fractions less than $\frac{1}{10}$ }
L = { fractions less than $\frac{1}{2}$ but greater than $\frac{1}{10}$ }.

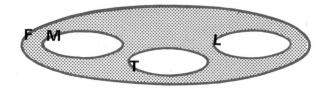

2 Which of the fractions should be written in the shaded part of the diagram?

3 Use the symbol ⊂ to show the sub-sets of F.

C

These are the names of the children who attend after-school clubs.

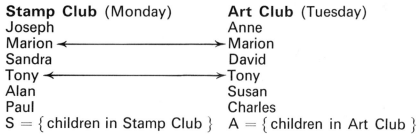

Stamp Club (Monday)	Art Club (Tuesday)
Joseph	Anne
Marion ⟷	Marion
Sandra	David
Tony ⟷	Tony
Alan	Susan
Paul	Charles

S = { children in Stamp Club }   A = { children in Art Club }

Look at the names of the members in each set.
You see that Marion and Tony are members of S and A.

When drawing the diagram, the rings for the sets are made to overlap or **to intersect**.
The names common to both sets are written in the overlap or **intersection**.
The diagrams you have been drawing have a special name.
They are called **Venn Diagrams** after the Rev. John Venn, a mathematician, who first used them in the nineteenth century.

Venn diagrams

A

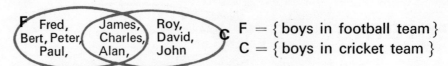

F = { boys in football team }
C = { boys in cricket team }

From the Venn diagram write the names of
1 all the boys who play in the school teams
2 the boys who play in the football team
3 the boys who play in the cricket team
4 the boys who play in both teams.

B

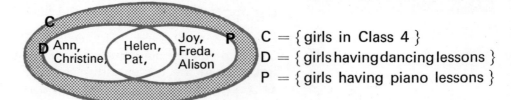

C = { girls in Class 4 }
D = { girls having dancing lessons }
P = { girls having piano lessons }

Answer the questions from the Venn diagram.
1 How many girls have **a** dancing lessons **b** piano lessons?
2 What can you learn about **a** Ann **b** Joy **c** Pat?
3 Does Freda have dancing lessons?
4 Betty is in Class 4. What do you know about her?

C

A number which will divide exactly into another number is called
a **Factor** of that number.

P = { the numbers between 1 and 19 of which 2 is a factor }
P = { 2, 4, 6, 8, 10, 12, 14, 16, 18 }
T = { the numbers between 2 and 19 of which 3 is a factor }
T = { 3, 6, 9, 12, 15, 18 }

1 Draw the Venn diagram and write in the numbers. In the overlap
or intersection show the numbers which have both 2 and 3 as factors.
2 Which is the smallest number divisible by 2 and 3?

D

1 5, 10, 15, 20, 25, 30, 35 ∈ { numbers divisible by 5 }
3, 6, 9, 12, 15, 18, 21, 24, 27, 30, 33 ∈ { numbers divisible by 3 }
Draw a Venn diagram to show that 15 is the smallest number which
is divisible by 5 and 3.

2 By drawing a Venn diagram show that, of all the numbers to 50,
36 is the only common multiple of 4 and 9.

E

N = { numbers 1 to 42 }
1 Describe **a** set F
b set S.
2 Why are the numbers
12, 24, 36 in the overlap
(the intersection)?

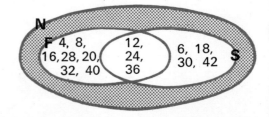

SETS — SYMBOLS TO REMEMBER

{ } means 'is a set of'   = means 'equal' or 'identical'

∅ means 'an empty set'   ≠ means 'not equal'

∈ means 'is a member of'   ⊂ means 'is a sub-set of'

∉ means 'is not a member of'   ⊄ means 'is not a sub-set of'

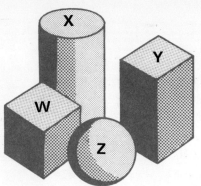

Surfaces and Solids
nets

A

1 Four different solids are shown in the picture. Which of these solids is
   a a cube   b a rectangular solid or cuboid
   c a cylinder   d a sphere?

2 Give two examples of common objects which are made in each of these shapes.

3 How many surfaces has a the cube and the cuboid
   b the cylinder   c the sphere?

B

1 cm^3

1 Obtain a 1 cm cube. What is the length of each side?
2 What is the area of
   a one surface of the cube   b all its surfaces?
3 Find the area of all the surfaces of
   a a 4 cm cube   b a 10 cm cube.

C

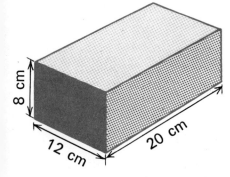

8 cm   12 cm   20 cm

1 What is a the length   b the width   c the height or thickness of the cuboid shown in the drawing?
2 Find the area of
   a its top and bottom
   b its long sides
   c its short sides.
3 What is the total area of the sides of the cuboid?

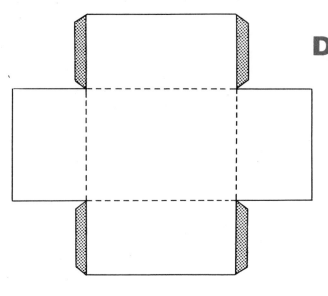

D

1 The diagram shows the net of a rectangular box with an open top. Find by measuring in cm
   a its length   b its width
   c its height.
2 What is the area in cm^2 of
   a the bottom
   b each of the long sides
   c each of the short sides?
3 On thin cardboard draw the net. Cut it out and fold it along the dotted lines. Use gum on the flaps to stick the box together. Keep the box you have made for later use.
4 Draw the net of a similar box with a top.

E

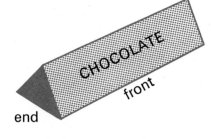

end   CHOCOLATE   front

This is a sketch of a box which contains chocolate. It measures 8 cm long, 3 cm wide and 2 cm high.
1 What is the shape of
   a the bottom   b the two ends   c the two front sides?
2 Draw full size the net of the box without any flaps.
3 Find the area in cm^2 of
   a the bottom   b each end   c each front side.

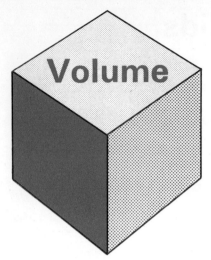

Volume

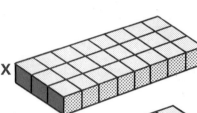

A

Get the box you have made (page 79, **D3**) and some centimetre cubes. The measurements of the box were length 4 cm, width 3 cm, height 2 cm.

1 Cover the bottom of the box with a layer of cubes. How many have you used?

2 How many such layers are required to fill the box?

3 How many cubes does the box hold?

The **amount of space** contained in the box is called its **volume** which is measured in centimetre cubes (cm³).

4 What is the volume in cm³ of your box?

5 If the box were 3 cm high
   a how many centimetre cubes would have been needed to fill it?
   b Write the volume of such a box.

B

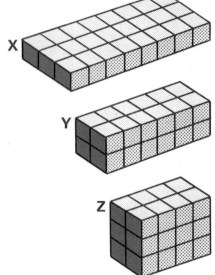

Cuboids X, Y and Z are made up of centimetre cubes.

1 a What is the length, width and height of cuboid X?
   b How many centimetre cubes are there in a row?
   c How many rows?
   d Find the volume of the cuboid in cm³.

2 Look at the cuboid Y.
   a Write its length, width and height.
   b How many cm³ are there in each layer?
   c How many layers?
   d Find the volume of cuboid Y.

3 Find
   a the dimensions of cuboid Z (its length, width, height)
   b its volume in cm³.

4 What have you discovered about the volume of X, Y and Z?

5 Get 36 centimetre cubes. Use them to make as many different cuboids as you can.
   Write
   a the dimensions (the length, width and height) of each
   b the volume of each.

6 Do this exercise again using 48 centimetre cubes.

C

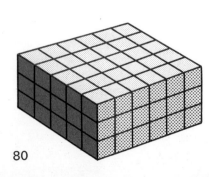

The diagram shows a cuboid made from centimetre cubes.

1 Write its dimensions.

2 Draw full size a plan of the bottom. Find its area in cm².

3 How many centimetre cubes fit into the bottom layer?

4 How many such layers are there?

5 Find in cm³ the volume of the cuboid.

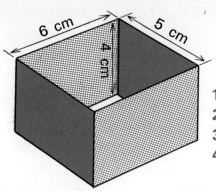

Volume capacity weight

A

1 The inside measurements of the box are shown. Write them.
2 How many cm³ can be fitted into the bottom layer?
3 How many layers are required to fill the box?
4 Write the volume of the box in cm³.

B

In the same way find the volume of each of the following.

	length	breadth	height		length	breadth	height
1	7 cm	5 cm	2 cm	2	9 cm	7 cm	4 cm
3	10 cm	6 cm	5 cm	4	13 cm	12 cm	8 cm
5	25 cm	20 cm	10 cm	6	37 cm	19 cm	6 cm
7	5 m	3·2 m	4 m	8	7 m	3·5 m	3 m

From the examples you have discovered that the volume of a cube or cuboid is found by multiplying together the length, breadth and height.
This is written for short as $V = l \times b \times h = lbh$ when
l is the number of units of length
b is the number of units of breadth and
h the number of units of height.

C

1 The outside measurements of a shoe box are 24 cm long, 18 cm wide and 9 cm high. Find how much space the box occupies.
2 a Which of these boxes takes up the most space?
   Box A   12 cm long   4 cm wide   3 cm high
   Box B   8 cm long   5 cm wide   3·5 cm high
   Box C   11 cm long   2·5 cm wide   4 cm high
   b Find the difference in cm³ between the largest box and the smallest box.
3 The inside measurements of a fish tank are 30 cm long, 22 cm wide and 20 cm high.
   Find
   a the volume of the tank
   b the volume of the water in the tank when it is filled to within 4 cm of the top.

D

Volume, capacity, weight
Turn back to page 33, Section **C**, and then answer these questions.
1 How many millilitres (ml) are equal to   a 1 cm³   b 1 000 cm³?
2 What liquid measure is equal to 1 000 cm³?
3 Write the weight of   a 1 cm³ of water   b 1 ml of water.
4 Write the weight of   a 1 000 cm³ of water   b 1 litre of water.
5 An empty ½ litre oil can is filled with water.
   What is
   a the volume of the water in cm³   b the weight of the water?
6 Find the weight in g of the water which will fill medicine bottles each holding
   a 300 ml   b 200 ml   c 150 ml   d 100 ml.
   Then write each answer in kg.
7 A glass tank measures 20 cm long, 20 cm wide and 10 cm high.
   Find
   a the volume of the tank in cm³
   b the number of litres of water the tank will hold
   c the weight of the water in kg.

Revision tests

A

In each example a letter stands for a number.
Find the value of the letter in each case.

1 $17 - x = 9$ **2** $a \times 12 = 0$ **3** $\frac{54}{6} = y$

4 $6 = 3 \cdot 7 + x$ **5** $8 \times 6 = 50 - y$ **6** $10z = 160$

7 $\frac{3}{4}$ of $d = 18$ **8** $5 (7 \cdot 5 + 2 \cdot 8) = z$ **9** $x = 3 (14 \cdot 6 - 3 \cdot 9)$

10 $\frac{81}{x} = 3 \times 3$ **11** $17x = 136$ **12** $6 \times a \times 4 = 168$

B

1 Write the value of the figure underlined in each of these numbers.
   a 19·4̲3   **b** 2̲0·961   **c** 3·06̲5   **d** 0·017̲

2 Write the following as decimal fractions.
   a 207 hundredths   **b** 805 tenths   **c** 1306 thousandths
   d $2\frac{1}{2}$ **e** $1\frac{3}{5}$ **f** $3\frac{3}{4}$ **g** $4\frac{3}{100}$ **h** $\frac{7}{20}$

Write these numbers

3 to the nearest 10   **a** 907   **b** 523   **c** 896

4 to the nearest 100   **a** 7 082   **b** 1 875   **c** 203·9

5 to the nearest tenth   **a** 3·06   **b** 12·88   **c** 9·97.

6 Arrange the following in order of size putting the largest first.
   a $\frac{2}{3}$ $\frac{5}{6}$ $\frac{3}{4}$ $\frac{1}{2}$ **b** $\frac{1}{4}$ $\frac{2}{5}$ $\frac{3}{10}$ $\frac{1}{2}$
   c 0·7   1·3   0·09   0·92   **d** 3·05   3·5   3·15   3·65

7 Multiply by 10   **a** 3·7   **b** 97   **c** £0·43   **d** £1·06.

8 Multiply by 100   **a** 530   **b** 7·8   **c** £0·25   **d** £0·09.

9 Divide by 10   **a** 68   **b** 2·5   **c** £2·80   **d** £0·30.

10 Divide by 100   **a** 702   **b** 41   **c** £2·00   **d** £27·00.

C

1 On squared paper draw a rectangle containing 100 squares.
Shade 10% of them red, 20% blue and 30% yellow.

2 What percentage is unshaded?

3 Write each of these percentages
   a as fractions in their lowest terms   **b** as decimal fractions.

D

1 20 bags of fertiliser weigh 250 kg.
Find the weight of   **a** 1 bag   **b** 7 bags.

2 In a concert hall the rows of seats are lettered A to N. If there are
26 seats in a row, find the total number of seats.

3 Draw this Venn diagram.
   $N = \{$ whole numbers between 1 and 10 $\}$
   $F = \{$ factors of 20 $\}$
   $G = \{$ factors of 18 $\}$
   Your diagram will show the only common factor of 18 and 20.
   What is it?

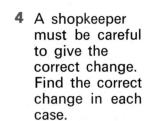

4 A shopkeeper must be careful to give the correct change. Find the correct change in each case.

	Price of goods	Notes or coins given
a	$37\frac{1}{2}$p	2 TENS, 4 FIVES
b	$81\frac{1}{2}$p	1 FIFTY, 1 TEN, 5 FIVES
c	£1·67	2 FIFTIES, 7 TENS
d	£3·63$\frac{1}{2}$	4 £1 notes

5 An electric train set costs £8·75.
   a Find the smallest number of notes and coins required to pay for it.
   b It can be paid for by 6 monthly payments each of £1·64. Find the
     difference between the cash price and the instalments.

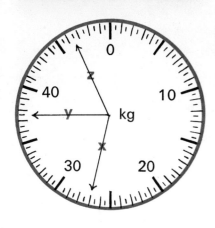

A

Write the following

1 as metres   **a** 1 460 mm   **b** 375 cm   **c** 850 mm

2 as kilogrammes **a** 250 g   **b** 900 g   **c** 2 500 g

3 as litres   **a** 1 500 cm³ **b** 700 ml   **c** 240 ml.

4 Find the cost of
  a 1·5 kg at 12p per ½ kg   **b** 200 g at 55p per ½ kg
  c 0·5 m at 75p per m   **d** 2·4 m at £1·50 per m
  e 550 g at 60p per ½ kg   **f** 6·5 litres at 38p per l.

5 The picture shows the dial on a weighing machine
suitable for finding the weights of children.
  a What is the greatest weight which can be
    measured on the scale?
  b What weight does each small division represent?
  c Find the weight of each child shown by the
    pointers **x**   **y**   **z.**
  d Find the average weight of the children.

6 Find the number of hours and minutes from
  a 10.30 a.m. to 1.05 p.m. **b** 9.15 p.m. to 2.05 a.m.
  c 1.13 p.m. to 3.07 p.m.   **d** 8.35 a.m. to 3.12 p.m.
  e 07.28 to 12.12   **f** 22.15 to 01.18.

7 **a** How many degrees are there in one complete
    turn?
  b What fraction of a complete turn is
    30°   60°   90°   180°   270°?

8 Find the value of
  a 50% of 270   **b** 25% of 94p   **c** 30% of £2·60.

B

1 This is a series of numbers.   3   6   9   12   15.
  a Multiply 37 by each number. What do you
    notice about the answers? Why do you think
    this is so?
  b Write the answers only to the following
    37 × 18   37 × 21.

2 Increase the following by 10%   **a** 65p   **b** £2·80.

3 Decrease the following by 20% **a** 75p   **b** £3·15.

4 Change these times to 24-hour clock times.
  a 3.15 a.m. **b** 9.25 a.m. **c** 4.37 p.m. **d** 9.05 p.m.

5 **a** How much is 6% of £1·00?
  b David puts a present of £17·00 in the bank.
    How much interest at 6% does he receive at the
    end of 1 year?

6 The diagram drawn to the scale 1 mm to 5 km
shows the course of a ship after leaving port.
  a Name the directions in which the ship sailed,
    then write them as bearings.
  By measuring find the actual distance
  b the ship sails in each direction
  c from port to the end of the course.

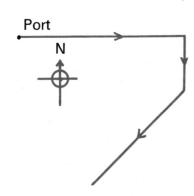

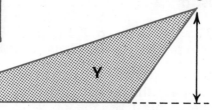

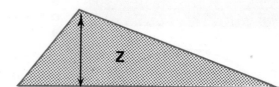

Revision tests

A

1 Find
  a the perimeter of a square the area of which is 64 cm²
  b the area of a rectangular room 5·6 m long and 3 m wide
  c the width of a rectangle the area of which is 44·8 cm² and the length 8 cm.

2 a Measure in mm the base and the height of triangle Y, triangle Z.
  b Which triangle has the larger area and by how many mm²?
  c Find the height of a triangle with the same area as Z if the base is half its length.

3 a Measure in mm the sides of the triangles A and B.
  b What kind of triangles are they?
  c How many of the smaller triangles can be cut from the larger one?

4 a To what measurement must a pair of compasses be set to draw circles of 17 cm and 56 mm in diameter?
  b Find the circumference of each circle to the nearest cm using $\pi = 3\cdot14$.

5 Find the radius of the largest semi-circle which can be cut from a rectangular sheet of paper which measures 11 cm long and 5 cm wide.

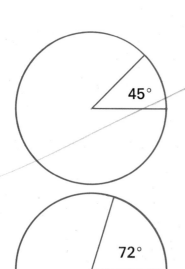

B

1 Draw a circle of 40 mm radius.

2 Use a protractor to draw an angle of 45° at the centre of the circle.

3 How many angles of 45° can be drawn at the centre?

4 Draw a regular octagon in the circle.

5 In another circle draw an angle of 72° at the centre. How many angles of 72° can be drawn at the centre?

6 Draw a regular pentagon in the circle.

7 Find the angle at the centre of a circle from which these regular shapes can be drawn in a circle
  a a hexagon b a decagon c a heptagon (7 sides).

8 a Find in cm³ the volume of this block of metal.
  b What would be the weight of an equal volume of water?
  c If the metal was 9 times heavier than water find the weight of the block in grammes. Write the answer in kg.

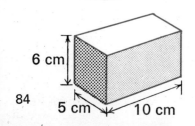

Facts to learn

Length

$$12 \text{ in} = 1 \text{ foot}$$
$$\left.\begin{array}{l} 3 \text{ ft} \\ 36 \text{ in} \end{array}\right\} = 1 \text{ yard}$$
$$22 \text{ yd} = 1 \text{ chain}$$
$$1\,760 \text{ yd} = 1 \text{ mile}$$

Weight

$$16 \text{ oz} = 1 \text{ lb}$$
$$14 \text{ lb} = 1 \text{ stone}$$
$$\left.\begin{array}{l} 8 \text{ st} \\ 112 \text{ lb} \end{array}\right\} = 1 \text{ cwt}$$
$$\left.\begin{array}{l} 20 \text{ cwt} \\ 2\,240 \text{ lb} \end{array}\right\} = 1 \text{ ton}$$

Capacity

$$4 \text{ gills} = 1 \text{ pint}$$
$$2 \text{ pt} = 1 \text{ quart}$$
$$\left.\begin{array}{l} 4 \text{ qt} \\ 8 \text{ pt} \end{array}\right\} = 1 \text{ gallon}$$

Imperial measures
length weight capacity

A

Cover the table. Write as quickly as you can the answers only to the following.

How many

1 inches in 1 ft?
2 inches in $\frac{1}{2}$ ft?
3 inches in $\frac{1}{4}$ ft?
4 inches in $\frac{3}{4}$ ft?
5 inches in 1 yd?
6 inches in $\frac{1}{2}$ yd?
7 inches in $\frac{1}{4}$ yd?
8 inches in $\frac{3}{4}$ yd?
9 yd in 1 chain?
10 yd in 1 mile?
11 yd in $\frac{1}{2}$ mile?
12 yd in $\frac{1}{4}$ mile?
13 yd in $\frac{3}{4}$ mile?
14 pt in 1 gal?
15 pt in $\frac{1}{2}$ gal?

16 oz in 1 lb?
17 oz in $\frac{1}{2}$ lb?
18 oz in $\frac{1}{4}$ lb?
19 oz in $\frac{3}{4}$ lb?
20 lb in 1 st?
21 lb in $\frac{1}{2}$ st?
22 lb in 1 cwt?
23 lb in $\frac{1}{2}$ cwt?
24 lb in $\frac{1}{4}$ cwt?
25 cwt in 1 ton?
26 cwt in $\frac{1}{2}$ ton?
27 cwt in $\frac{1}{4}$ ton?
28 cwt in $\frac{3}{4}$ ton?
29 lb in 1 ton?
30 lb in $\frac{1}{2}$ ton?

B

Change

	a	b	c
1	47 in to ft and in	38 ft to yd and ft	73 in to yd and in
2	5 ft 9 in to in	8 yd 2 ft to ft	3 yd 21 in to in
3	49 oz to lb and oz	69 lb to st and lb	52 cwt to tons and cwt
4	5 lb 3 oz to oz	5 st 13 lb to lb	3 tons 17 cwt to cwt
5	31 pt to gal and pt	6 gal 5 pt to pt	4 qt 1 pt to pt.

C

1 Find the total of
 a 3 ft 11 in,   2 ft 8$\frac{1}{2}$ in,   5 ft 9 in
 b 1 lb 10 oz,   3 lb 5$\frac{1}{2}$ oz,   2 lb 4$\frac{1}{2}$ oz
 c 3 st 11 lb,   4 st 9 lb,   4 st 5 lb
 d 2 gal 3 pt,   4 gal 7 pt,   1 gal 2 pt.

2 Find the difference between
 a 5$\frac{1}{2}$ gal and 3 gal 5 pt   b 19 lb 6 oz and 7 lb 12 oz
 c 15 ft 7$\frac{1}{2}$ in and 7 ft 10 in.

3 a 4 lb 12 oz × 4   b 4 ft 6 in × 8   c 2 tons 7 cwt × 10

4 a 15 ft 9 in ÷ 7   b 14 lb 10 oz ÷ 6   c 22 gal 4 pt ÷ 9

5 a $\frac{1}{5}$ of 4 yd 16 in   b $\frac{1}{4}$ of 25 lb   c $\frac{3}{4}$ of 3 ft 6 in

6 Find the answers to the following as indicated.
 a 15 lb ÷ 9 (to the nearest oz) b 3 cwt ÷ 13 (to the nearest lb)
 c 1 mile ÷ 7 (to the nearest yd) d 14 ft 7 in ÷ 8 (to the nearest in)
 e 5$\frac{1}{2}$ gal ÷ 6 (to the nearest pt) f 4 yd 2 ft 7 in ÷ 3 (to the nearest in)

Imperial measures
length
weight
capacity

A

Find the cost of

1 $\frac{3}{4}$ lb at $1\frac{1}{2}$p per oz
2 $2\frac{1}{2}$ lb at 36p per lb
3 $1\frac{1}{4}$ yd at 92p per yd
4 $4\frac{1}{3}$ yd at £3·75 per yd
5 $3\frac{1}{2}$ lb at 48p per lb
6 $1\frac{1}{2}$ tons at £10·80 per ton
7 1 gal 6 pt at 14p per pt
8 7 pt at 44p per gal
9 12 oz at 62p per lb
10 2 lb 4 oz at 28p per lb.

B

Copy this ready reckoner and fill in the missing items.

Cost per lb	8p	16p	24p			48p		64p
Cost per oz				2p	$2\frac{1}{2}$p		$3\frac{1}{2}$p	

First check your ready reckoner, then use it to find the following costs.

1 3 oz at 16p per lb
2 7 oz at 32p per lb
3 11 oz at 64p per lb
4 1 lb 5 oz at 8p per lb
5 1 lb 13 oz at 48p per lb
6 1 lb 2 oz at 56p per lb
7 2 lb 9 oz at 24p per lb
8 3 lb 10 oz at 32p per lb
9 4 lb 10 oz at 48p per lb
10 Lamb chops cost 32p per lb.
   a How much per oz?
   b Find the weight of chops in lb and oz for 54p.
11 Apples cost 8p per lb. Find the weight of apples in lb and oz for 23p.
12 Beef costs 56p per lb. Find the weight of beef in lb and oz for 70p.

C

1 A coil of rope is cut into 6 equal pieces each $27\frac{1}{4}$ yd long. If 9 in of rope was left over, find the total length of the coil.

2 There are 9 gal 3 pt of oil in a tank which holds 16 gal.
   a How much oil is required to fill the tank?
   b How many 2 pt tins can be filled from a full tank?

3 The table shows the heights and weights of 5 children.

Name	Height	Weight
Sally	4′ 11″	5 st $1\frac{1}{2}$ lb
Peter	5′ $3\frac{1}{2}$″	5 st 5 lb
Joan	5′ 2″	5 st $3\frac{1}{2}$ lb
Terry	4′ 10″	5 st 3 lb
Richard	5′ $1\frac{1}{2}$″	5 st $9\frac{1}{2}$ lb

   a What is the difference in height between the tallest and the shortest child?
   b What is the difference in weight between the heaviest and the lightest child?
   c Find the average height to the nearest inch.
   d Find the average weight to the nearest lb.

4 A 2-gallon jar when filled with water weighed 23 lb 7 oz. If 1 pint of water weighs $1\frac{1}{4}$ lb find the weight of the jar.

5 Find a the missing measurements marked x and y on the drawing
        b the perimeter of the shape in ft and in.

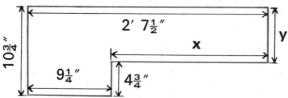

D

1 1 kg = 2·2 lb approx.
   How many lb in a 5 kg   b 12 kg?
2 1 litre = $1\frac{3}{4}$ pt approx.
   How many pt in a 10 litres   b 40 litres?
3 1 metre = 39·4 in approx.
   How many ft and in in a 3 m   b 7 m?
4 1 km = $\frac{5}{8}$ mile approx.
   How many miles in a 24 km   b 100 km?

86

Imperial measures ruler

A

1 Measure each of these lines accurately with a ruler marked in $\frac{1}{2}''$ $\frac{1}{4}''$ $\frac{1}{8}''$ $\frac{1}{10}''$.

a _____
b _____
c _____
d _____
e _____

2 Now write each measurement to the nearest inch.

3 Change to an improper fraction the measurement of each line   **a   b   c   d   e.**

B

Twelfths

1 On this ruler, each inch is divided into 12 equal parts. Count them.

2 How many twelfths in
   a 2″ **b** 5″ **c** $\frac{1}{2}''$ **d** $1\frac{1}{2}''$ **e** $2\frac{1}{4}''$ **f** $3\frac{3}{4}''$?

3 Find twelfths on your ruler and draw lines of the following lengths
   a $1\frac{1}{12}''$ **b** $3\frac{5}{12}''$ **c** $5\frac{7}{12}''$ **d** $2\frac{11}{12}''$ **e** $4\frac{3}{4}''$ **f** $6\frac{2}{3}''$.

4 Measure each of these lines using inches and twelfths on your ruler.

a _____
b _____
c _____
d _____

5 Now write each measurement to the nearest $\frac{1}{2}$ inch.

6 Change to an improper fraction the measurement of each line   **a   b   c   d.**

C

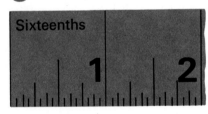

Sixteenths

1 On this ruler each inch is divided into 16 equal parts. Count them.

2 How many sixteenths in
   a 2″ **b** 6″ **c** $\frac{1}{2}''$ **d** $1\frac{1}{4}''$ **e** $1\frac{3}{8}''$ **f** $3\frac{7}{16}''$?

3 Find sixteenths on your ruler and draw lines of the following lengths
   a $2\frac{9}{16}''$ **b** $4\frac{3}{16}''$ **c** $6\frac{11}{16}''$ **d** $5\frac{5}{16}''$ **e** $3\frac{3}{4}''$ **f** $2\frac{7}{8}''$.

4 Measure each of these lines using inches and sixteenths on your ruler.

a _____
b _____
c _____
d _____

5 Now write each measurement to the nearest $\frac{1}{2}$ inch.

6 Change to an improper fraction the measurement of each line   **a   b   c   d.**

D

1 What fraction of 1 foot is   **a** 1″ **b** 5″ **c** 7″ **d** 11″?

2 Write in feet only   **a** 3′ 5″ **b** 2′ 7″ **c** 1 ft 11″.

3 Write as ft and in   **a** $2\frac{3}{4}$ ft   **b** $4\frac{1}{12}$ ft   **c** $3\frac{5}{12}$ ft   **d** $1\frac{7}{12}$ ft.

4 What fraction of 1 lb is   **a** 1 oz   **b** 3 oz   **c** 9 oz   **d** 15 oz?

5 Write in lb only   **a** 1 lb 7 oz   **b** 2 lb 11 oz   **c** 4 lb 3 oz.

6 Write in lb and oz   **a** $3\frac{1}{4}$ lb   **b** $2\frac{5}{16}$ lb   **c** $5\frac{13}{16}$ lb.

87

Imperial measures

ruler measurements

Turn to page 19. Work again from Section **B** the addition examples **1a, 2a, 3a, 4a**.

A

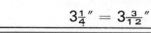 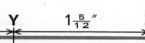

X $\quad\quad\quad 3\frac{1}{4}'' = 3\frac{3}{12}''\quad\quad\quad$ Y $\quad 1\frac{5}{12}''\quad$ Z

1 a Write the length of the line XY, the line YZ.
  b The length of the line XZ $= 3\frac{3}{12}'' + 1\frac{5}{12}''$.
    Find the length of XZ. Check the answer by measuring.

2 Find the total length of
  a $1\frac{1}{4}'' + 2\frac{1}{2}''$ b $3\frac{1}{2}'' + 1\frac{3}{8}''$ c $3\frac{3}{10}'' + 1\frac{1}{2}''$.
  Check the answers by drawing lines and measuring the total length.

B

Example

$1\frac{3}{4} + 1\frac{2}{3}$
$1\frac{9}{12} + 1\frac{8}{12}$
$= 2 + \frac{9}{12} + \frac{8}{12}$
$= 2 + \frac{17}{12} \left(\frac{17}{12} = 1\frac{5}{12}\right)$
$= 3\frac{5}{12}$

Look at this example.
To **add** measurements with **mixed numbers** first change the fractions to the **same name** (the **same denominator**).
Add the **whole numbers,** then the **fractions.**
Work the following.

	a	b	c	d
1	$1\frac{1}{4} + 2\frac{1}{3}$	$1\frac{3}{4} + 4\frac{1}{8}$	$2\frac{1}{6} + 2\frac{2}{3}$	$4\frac{1}{2} + 3\frac{1}{12}$
2	$6\frac{1}{2} + 2\frac{3}{16}$	$2\frac{7}{10} + 2\frac{1}{5}$	$4\frac{9}{16} + 1\frac{3}{8}$	$2\frac{3}{10} + 1\frac{2}{5}$
3	$2\frac{7}{10} + 1\frac{4}{5}$	$5\frac{1}{2} + 3\frac{9}{16}$	$1\frac{5}{8} + 2\frac{3}{4}$	$6\frac{5}{6} + 1\frac{3}{4}$
4	$3\frac{1}{2} + 4\frac{5}{8}$	$3\frac{11}{12} + 2\frac{2}{3}$	$1\frac{1}{2} + 1\frac{2}{3}$	$4\frac{4}{5} + 4\frac{1}{2}$

C

1 a Write the length of the line AB, the line BC.
  b The length of the line AC $= 4\frac{6}{8}'' - 1\frac{3}{8}''$.
    Find the length of AC. Check the answer by measuring.

A $\quad\quad\quad 4\frac{3}{4}'' = 4\frac{6}{8}''\quad\quad\quad$ C $\quad 1\frac{3}{8}''\quad$ B

2 Find the length of
  a $2\frac{1}{2}'' - 1\frac{3}{8}''$ b $3\frac{11}{12}'' - 1\frac{1}{2}''$ c $4\frac{1}{2}'' - 2\frac{1}{10}''$.
  Check the answers by drawing lines and measuring.

Look at this example.
To subtract measurements with **mixed numbers** first change the fractions to the **same name** (the **same denominator**).
Subtract the whole numbers, then the fractions.

Work the following.

3 $2\frac{3}{5} - 1\frac{1}{10}$ 4 $5\frac{11}{12} - \frac{2}{3}$ 5 $3\frac{5}{6} - 2\frac{2}{3}$ 6 $7\frac{7}{8} - 3\frac{1}{2}$

Example

$3\frac{3}{4} - 1\frac{5}{16}$
$= 3\frac{12}{16} - 1\frac{5}{16}$
$= 2\frac{12}{16} - \frac{5}{16}$
$= 2\frac{7}{16}$

D

Turn to page 19. Work again from Section **E** the subtraction examples **1, 3, 5, 7, 9.**
Look at the example and then work the following.

1 $2\frac{3}{10} - 1\frac{4}{5}$ 2 $5\frac{1}{4} - 2\frac{3}{8}$
3 $3\frac{1}{12} - 2\frac{3}{4}$ 4 $10\frac{1}{5} - 3\frac{7}{10}$
5 $2\frac{3}{8} - 1\frac{13}{16}$ 6 $5\frac{1}{2} - 1\frac{9}{10}$
7 $3\frac{1}{8} - 2\frac{3}{4}$ 8 $4\frac{5}{6} - 1\frac{11}{12}$
9 $8\frac{2}{5} - 3\frac{1}{2}$ 10 $2\frac{1}{4} - 1\frac{5}{8}$
11 $1\frac{1}{3} - \frac{5}{12}$ 12 $2\frac{3}{8} - 1\frac{9}{16}$
13 $7\frac{7}{12} - 2\frac{3}{4}$ 14 $1\frac{13}{16} - \frac{7}{8}$

Example

$4\frac{1}{4} - 2\frac{2}{3}$
$= 2\frac{1}{4} - \frac{2}{3}$
$= 2\frac{3}{12} - \frac{8}{12}$
It is impossible to take $\frac{8}{12}$ from $\frac{3}{12}$ so change one of the whole ones to twelfths and add it to $\frac{3}{12}$ making $\frac{15}{12}$
$= 1\frac{15}{12} - \frac{8}{12}$
$= 1\frac{7}{12}$

Imperial measures

A

1 Draw a line AB which is $5\frac{5}{12}''$ long.
From B measure back to C, a distance of $3\frac{3}{4}''$.
a Find by calculation the length of AC.
b Check your answer by measuring AC using twelfths on the ruler.

2 From the sum of $2\frac{7}{8}$ and $1\frac{3}{4}$ take $3\frac{1}{2}$.

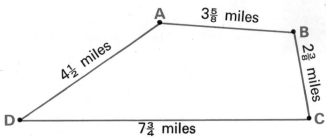

3 A, B, C, D are the positions of four villages.
Find the distance from B to D
a via A   b via C.

4 Which is the shorter distance and by how many miles?

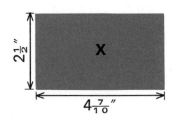
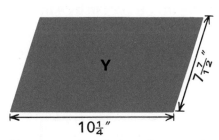
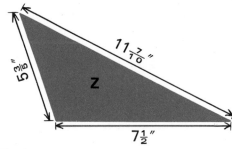

5 Name each of the shapes above. Find
a the difference between the longest and shortest sides of each shape
b the perimeter of each shape.

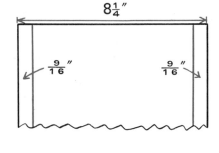

6 A boy marks out his paper with two equal margins as shown.
a Find the width left for writing.
b What distance from the margin line is the middle point of the width?

7 Find the radius of a circle the diameter of which is
a $2\frac{3}{4}''$ b $4\frac{3}{8}''$ c $7\frac{6}{10}''$.

8 Find the diameter of a circle the radius of which is
a $1\frac{7}{8}''$ b $3\frac{7}{10}''$ c $2\frac{7}{12}''$.

B

1 Find the actual distances represented by each of the following lines.
a Scale 1 inch to 1 foot (use twelfths on the ruler).

_____

b Scale 1 inch to 16 yards (use sixteenths on the ruler).

_____

c Scale 1 inch to 10 miles (use tenths on the ruler).

_____

2 A plan is drawn to a scale $\frac{1}{8}$ in to 1 ft.
Find the length on the plan which represents
a 17 ft   b 39 ft   c 6 in   d 7 ft 6 in.

3 On a map drawn to a scale of 1 in to 4 miles,
A, B and C are in a straight line.
A is $3\frac{3}{4}''$ distant from B and C is $5\frac{3}{8}''$ distant from B.
Find the actual distance from A to B, from B to C, from A to C.

Imperial measures
area
squares
rectangles

Turn to pages 50 and 51. Make sure you understand these formulae.

$A = l \times b = lb$	$l = A \div b = \dfrac{A}{b}$	$b = A \div l = \dfrac{A}{l}$

Area is measured in **square inches (in²)**, **square feet (ft²)**, **square yards (yd²)**.

A

1 Find the area of each of these rectangles. Look carefully at the measurements of the length and width before writing the answer.

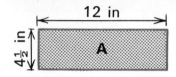

12 in — 4½ in — A

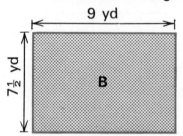

9 yd — 7½ yd — B

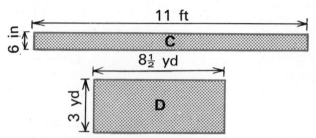

11 ft — 6 in — C
8½ yd — 3 yd — D

Find the area of each of the following squares or rectangles.

2 a 8 in by 6½ in   b 10¾ in by 4 in   c 16 in square
3 a 7½ ft by 5 ft   b 17 ft square   c 45 yd by 30 yd
4 a 1′ 6″ by 15″   b 13′ 6″ by 12′   c 2½ yd square

B

Find the missing measurement in each of the following squares and rectangles.

1

	Area	Length	Breadth
a	27 sq in	3″	
b	64 sq in		8″
c		6½″	9″

2

	Area	Length	Breadth
a	75 sq in		15″
b	180 sq ft	12′	
c		8′ 3″	7′

3

	Area	Length	Breadth
a	27½ sq yd	5 yd	
b	225 sq ft		15′
c		8¾″	5″

C

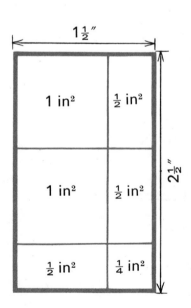

1½″ — 2½″ — 1 in² | ½ in² | 1 in² | ½ in² | ½ in² | ¼ in²

1 Look at the rectangle. What is **a** its length   **b** its breadth?
2 Its surface has been marked out in square inches and parts of a square inch. By counting, find the area of the rectangle.
3 Using the formula **A** = **l** × **b**, fill in the numbers for the letters **l** and **b** and write $A = (2\frac{1}{2} \times 1\frac{1}{2})in^2$.
   a Change the mixed numbers to improper fractions $A = (\frac{5}{2} \times \frac{3}{2})\ in^2$.
   b Multiply numerators, multiply denominators $A = \frac{15}{4} = 3\frac{3}{4}\ in^2$.
   This answer should be the same as that in No. 2.
4 Now find the areas of the following squares and rectangles in two ways
   a by drawing the shape and marking it out
   b by using the formula **A** = **l** × **b**.
   a square of 1½″ sides   a rectangle 3½″ by ½″
   a square of 3½″ sides   a rectangle 4½″ by 2½″

D

1 How many in² are there in a 12 inch square?
   Write 1 ft² = ☐ in².
2 Find the area of **a** a table top 3½ ft square
   b a plank of wood 3′ 6″ long and 6″ wide.
   Write the answers in ft² and in².
3 How many ft² are there in a square of 3 ft sides?
   Write 1 yd² = ☐ ft².

Imperial measures
area and volume

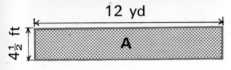

12 yd

$4\frac{1}{2}$ ft

A

A

1 Find the perimeter of **a** a square of 18″ side
  b a rectangle $9\frac{1}{2}$″ long and 7″ wide.

2 Find the area of
  a the square **b** the rectangle in **1a** and **b**.

3 Find the perimeter and the area of
  a a room 10′ 6″ by 9′ **b** a carpet $3\frac{1}{2}$ yd by $2\frac{1}{2}$ yd.

4 Find
  a the perimeter **b** the area of each of shapes A, B and C.

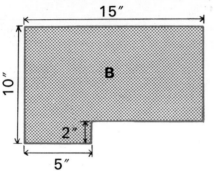

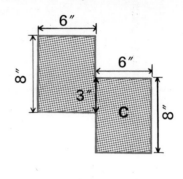

B

Turn to page 67. You discovered

Area of a triangle $= \frac{1}{2}$ **(base** $\times$ **height)** $= \dfrac{\textbf{base} \times \textbf{height}}{2}$

1 **a** Find the area of these triangles.

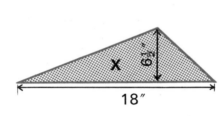

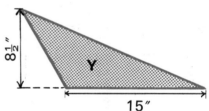

b Which triangle has the greater area and by how many in²?

2 Find the area of each of these triangles.
  a base 24″ height $10\frac{1}{2}$″ **b** base $4\frac{1}{2}$ ft height 3 ft
  c base 19″ height $7\frac{1}{2}$″ **d** base 5′ 6″ height 2′

C

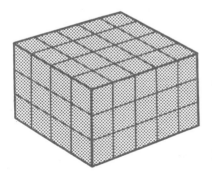

Turn to pages 80 and 81. Make sure you understand the formula for finding the volume of a cube or cuboid.

$\textbf{V} = \textbf{l} \times \textbf{b} \times \textbf{h} = \textbf{lbh}$

1 Volume is measured in inch cubes (in³), foot cubes (ft³) or yard cubes (yd³).
  What is the length, breadth and height of each of these cubes?

2 The drawing shows a cuboid made from inch cubes.
  a Write its dimensions.
  b How many in³ are there in the bottom layer?
  c How many layers?
  d Find the volume of the cuboid.

3 Find the volume of cubes of
  a 14″ side **b** 1′ 6″ side **c** 20″ side.

4 Look carefully at the measurements in the table.
  Find the volume of each cuboid.

	length	breadth	height
a	9″	8″	$7\frac{1}{2}$″
b	$5\frac{1}{2}$″	$3\frac{1}{2}$″	4″
c	1′ 3″	12″	5″
d	7′ 6″	6′ 6″	8′

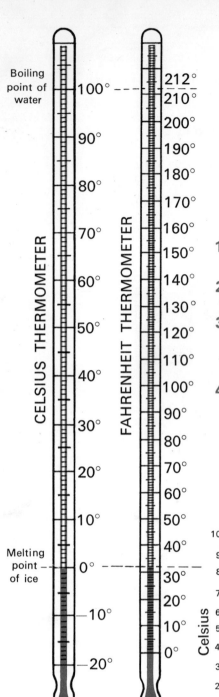

Temperature
Celsius and Fahrenheit

A

You have already practised using a **Celsius** thermometer. Sometimes, however, temperatures are measured on a **Fahrenheit** thermometer which has a **different scale**.

In the picture both thermometers are drawn side by side. By placing a ruler across the scales temperatures can be easily changed from one to the other.

1 On the Celsius scale boiling point is 100°, freezing point is 0°. Find each of these temperatures on the Fahrenheit scale.

2 How many degrees are there between boiling point and freezing point on
   a the Celsius scale   b the Fahrenheit scale?

3 By placing your ruler across the two scales change to Fahrenheit temperatures
   a 0°C   b 5°C   c 15°C   d 29°C
   e 35°C   f 63°C   g −5°C   h −12°C.

4 In the same way change to Celsius temperatures
   a 212°F   b 175°F   c 140°F   d 104°F
   e 70°F   f 50°F   g 15°F   h 0°F.

B

Conversion graph
(Celsius to Fahrenheit or Fahrenheit to Celsius)

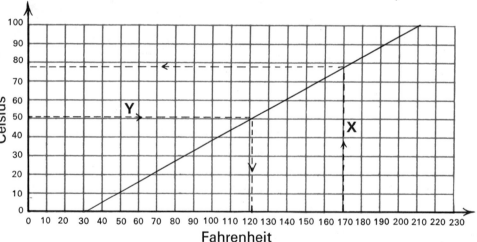

Copy the graph on squared paper.
Make the divisions as large as possible.

1 Draw the vertical axis and mark the Celsius scale.

2 Draw the horizontal axis and mark the Fahrenheit scale.

3 Plot the points a 0°C and 32°F b 100°C and 212°F
   Join these points by a straight line.
   Use your graph to convert one scale to the other.
   Two examples are done for you.
   Example marked X   170°F = 77°C
   Example marked Y   50°C = 122°F.

4 Change to Fahrenheit scale
   a 10°C   b 20°C   c 30°C   d 40°C   e 60°C   f 90°C.

5 Change to Celsius scale
   a 40°F   b 60°F   c 80°F   d 120°F   e 160°F   f 200°F.

6 Use a Celsius thermometer to take the temperature of the air
   a inside the classroom   b in the playground.
   Use your graph to convert these temperatures to the Fahrenheit scale.